Official Organ of the European Society for Paediatric Endocrinology

Hormone Research

International Journal of Experimental and Clinical Endocrinology

Founded 1970 as 'Hormones' by M. Marois

KARGER

Printed in Switzerland
on acid-free paper by
Thür AG Offsetdruck, Pratteln

Appears monthly:
2 volumes per year (12 issues)

Hormone Research

'Hormone Research' aims to provide rapid publication of clinical investigations and case reports, as well as experimental and methodological work related to endocrinology. Correspondence is welcome, but publication of letters is at the discretion of the editors.

Submission: Only original papers written in English are considered and should be sent to:

S. Karger AG
Editorial Dept. 'Hormone Research'
P.O. Box
CH-4009 Basel (Switzerland)

Manuscripts should be submitted in triplicate (with three sets of illustrations of which one is an original), typewritten double-spaced on one side of the paper, with a wide margin.

Conditions: All manuscripts are subject to editorial review. Manuscripts are received with the explicit understanding that they are not under simultaneous consideration by any other publication. Submission of an article for publication implies the transfer of the copyright from the author to the publisher upon acceptance. Accepted papers become the permanent property of 'Hormone Research' and may not be reproduced by any means, in whole or in part, without the written consent of the publisher. It is the author's responsibility to obtain permission to reproduce illustrations, tables, etc. from other publications.

Arrangement:
Title page: The first page of each paper should indicate the title (main title underlined), the authors' names, and the institute where the work was conducted. A short title for use as running head as well as the full address of the author to whom correspondence should be sent are also required.
Full address: The exact postal address complete with postal code must be given at the bottom of the title page. Please also supply Phone, Fax, and E-mail numbers.
Key words: For indexing purposes, a list of 3-10 key words in English is essential.
Abstract: Each paper needs an abstract of up to 10 lines.
Small type: Paragraphs which can or must be set in smaller type (case histories, test methods, etc.) should be indicated with a 'p' (petit) in the margin on the left-hand side.
Footnotes: Avoid footnotes. When essential, they are numbered consecutively and typed at the foot of the appropriate page.
Tables and illustrations: Tables and illustrations (both numbered in Arabic numerals) should be prepared on separate sheets. Tables require a heading and figures a legend, also prepared on a separate sheet. For the reproduction of illustrations, only good drawings and original photographs can be accepted; negatives or photocopies cannot be used. Due to technical reasons, figures with a screen background should not be submitted. When possible, group several illustrations on one block for reproduction (max. size 181 × 223 mm) or provide crop marks. On the back of each illustration, indicate its number, the author's name, and 'top' with a soft pencil.

Color illustrations: Up to 6 color illustrations per page can now be integrated with the text at the reduced price of CHF 600.-/USD 545.00 per page. Color illustrations are reproduced at the author's expense.

References: In the text identify references by Arabic numerals [in square brackets]. Material submitted for publication but not yet accepted should be noted as 'unpublished data' and not be included in the reference list. The list of references should include only those publications which are cited in the text. Do not alphabetize; number references in the order in which they are first mentioned in the text. The surnames of the authors followed by initials should be given. There should be no punctuation other than a comma to separate the authors. Cite all authors, 'et al' is not sufficient. Abbreviate journal names according to the Index Medicus system. (Also see International Committee of Medical Journal Editors: Uniform requirements for manuscripts submitted to biomedical journals. Br Med J 1988;296:401-405.)
Examples
(a) Papers published in periodicals: Kauffman HF, van der Heide S, Beaumont F, Blok H, de Vries K: Class-specific antibody determination against *Aspergillus fumigatus* by means of the enzyme-linked immunosorbent assay. III. Comparative study: IgG, IgA, IgM ELISA titers, precipitating antibodies and IgE binding after fractionation of the antigen. Int Arch Allergy Appl Immunol 1986;80:300-306.
(b) Monographs: Matthews DE, Farewell VT: Using and Understanding Medical Statistics. Basel, Karger, 1985.
(c) Edited books: Hardy WD Jr, Essex M: FeLV-induced feline acquired immune deficiency syndrome: A model for human AIDS; in Klein E (ed): Acquired Immunodeficiency Syndrome. Prog Allergy. Basel, Karger, 1986, vol 37, pp 353-376.
Further details: Authors may consult our leaflet 'Rules for the Preparation of Manuscripts', available on request from the Publisher, S. Karger AG, P.O. Box, CH-4009 Basel (Switzerland).

Page charges: There is no page charge for papers of 4 or fewer printed pages (including tables, illustrations and references). Each additional complete or partial page is charged to the author at CHF 510.-/USD 400.00. The allotted size of a paper is equal to approx. 12 manuscript pages (including tables, illustrations and references). As the page charges are waived in any case for the first paper each year to be published by a member of ESPE, authors who are members of the ESPE should so indicate on the title page.

Galley proofs: Unless indicated otherwise, galley proofs are sent to the first-named author and should be returned with the least possible delay. Alterations made in galley proofs, other than the correction of printer's errors, are charged to the author. No page proofs are supplied.

Reprints: Order forms and a price list are sent with the galley proofs. Orders submitted after the issue is printed are subject to considerably higher prices. Allow five weeks from date of publication for delivery of reprints.

KARGER

General Information

Hormone Research

Publication data: 'Hormone Research' is published 12 times annually. Volumes 43–44 with 6 issues each appear in 1995
Subscription rates: Subscriptions run for a full calendar year. Prices are given per volume.
Personal subscription:

Subscription	*+ postage and handling*
SFr. 300.30	+ SFr. 17.90 Europe
SFr. 300.30	+ SFr. 27.30 Overseas
US$ 231.00	+ US$ 16.80

(Must be in the name of, billed to, and paid by an individual. Order must be marked 'personal subscription'.)
Institutional subscription:

Subscription	*+ postage and handling*
SFr. 429.–	+ SFr. 25.50 Europe
SFr. 429.–	+ SFr. 39.– Overseas
US$ 330.00	+ US$ 24.00

Airmail postage: SFr. 36.–, US$ 30.00. (Extra per volume.)
Discount subscription prices:
European Society of Paediatric Endocrinology. Lawson Wilkins Society; ISGD; Schweizerische Gesellschaft für Endokrinologie.
Advertising: Correspondence and rate requests should be addressed to the Publisher.

Microforms: Please direct all enquiries and orders to:
University Microfilms Inc.
300 North Zeeb Road, P.O. Box 1346
Ann Arbor, MI 48106-1346 (USA)
Subscription orders: Orders can be placed at agencies, bookstores, directly with the Publisher S. Karger AG, P.O. Box, CH–4009 Basel (Switzerland)
or with any of the following distributors:
Deutschland: S. Karger GmbH, Postfach
D–79095 Freiburg
France: Librairie Luginbühl
36, bd de Latour-Maubourg, F–75007 Paris
Great Britain, Ireland: S. Karger Publishers
11 Ravenscroft Road
Beckenham, Kent BR3 4TP
USA: S. Karger Publishers, Inc.
26 West Avon Road, P.O. Box 529
Farmington, CT 06085
India, Bangladesh, Sri Lanka:
S. Karger India Subscription Agency
B–5/132 Safdarjung Enclave
New Delhi-110029
Japan: Katakura Libri, Inc.
Take 1 Bldg, 2nd Floor, 4-5-7 Yushima
Bunkyo-ku, Tokyo 113
Change of address: Both old and new address should be sent to the subscription source.

Single issues and back volumes: Information can be obtained through the Publisher.
Bibliographic indices: This journal is regularly listed in bibliographic services, including *Current Contents®*.
Book Reviews: Books are accepted for review by special agreement.
Copying: This journal has been registered with the Copyright Clearance Center (CCC), as indicated by the code appearing on the first page of each article. For readers in the US, this code signals consent for copying of articles for personal or internal use, or for the personal or internal use of specific clients, provided that the stated fee is paid per copy directly to Copyright Clearance Center, Inc., 222 Rosewood Drive, Danvers, MA 01923 (USA). A copy of the first page of the article must accompany payment. Consent does not extend to copying for general distribution, for promotion, for creating new works, or for resale. In these cases, specific written permission must be obtained from the copyright owner, S. Karger AG, P.O. Box, CH–4009 Basel (Switzerland).

Guidelines for Disk Submission

Karger Publishers welcomes the submission of manuscripts on disk.
Please observe the 'Instructions to Authors' as well as the following guidelines:

Technical Data

Storage medium is a 3.5- or 5.25-inch disk. **The preferred word processing package is Word for Windows®.** Other commonly used PC text programs are also accepted, as well as ASCII formats. MacIntosh® is also acceptable. However **1,44 Byte HD** in ASCII is preferred.

Handling

All manuscripts accepted for Karger publications are subject to copy-editing. It is therefore essential that you send us a **double-spaced printout** of your manuscript which exactly matches the disk.
Any subsequent revised versions should be clearly marked as such, both on the disk and on the manuscript.

The copy-edited manuscript and the disk will be returned to you along with the galley proofs. Please be sure to **check the galleys against the manuscript** in which changes introduced by the copy-editor and/or questions will be marked.
Please **label** your disk with (a) the title of the journal or book for which it is submitted, (b) a short version of the paper title, (c) your name, (d) the software and hardware used and (e) the file names contained.
Do not place more than one article in a file.

Text Arrangement

Please follow the general rules about style given in the 'Instructions to Authors' or 'Rules for the Preparation of Manuscripts'.
Be sure to use the same type style for headings of the same ranking throughout the text. Be consistent and stick to the same work pattern throughout the manuscript.
Do not use automatic page numbering. Number the pages by hand.
Text and headings should be ranged left throughout, without word breaks. **Do not center, space, or use block letters** for text and headings (except for abbreviations). Do not split words (neither by hand nor automatically). **Do not use indentations** anywhere in the manuscript, including the references.
To mark the beginning of a paragraph, leave one blank line between paragraphs. Do not use a hyphenating program. Use the possibilities of your text program to emphasize text parts, e.g. **Bold,** *Italic,* $_{\text{Sub}}$- and ^{Superscript}.
Insert one (and only one) space after each interpunctuation mark – e.g. at the end of sentences and with reference numbers, i.e. [22, 27, 35].
When using **special characters** not available on a standard keyboard (including Greek characters) please do so by a readily identifiable code, and use the same code throughout the manuscript. Supply a **complete list of the codes used** at the beginning of the text or include a 'Read-me-file'.
Do not use spaces in tables; use the tabulator. Operation data within tables (e.g. 15.1 ± 7.23) should be formatted in one column, not split into two.
Illustrations and mathematical formulae will be handled conventionally.

KARGER

Growth and Growth Disorders: An Update

Diagnostic and Therapeutic Aspects

Guest Editor
G. Saggese, Pisa

52 figures and 13 tables, 1995

KARGER

Hormone Research
Vol. 44, Supplement 3, 1995

S. Karger
Medical and Scientific Publishers
Basel · Freiburg · Paris · London
New York · New Delhi · Bangkok
Singapore · Tokyo · Sydney

Drug Dosage
The authors and the publisher have exerted every effort to ensure that drug selection and dosage set forth in this text are in accord with current recommendations and practice at the time of publication. However, in view of ongoing research, changes in government regulations, and the constant flow of information relating to drug therapy and drug reactions, the reader is urged to check the package insert for each drug for any change in indications and dosage and for added warnings and precautions. This is particularly important when the recommended agent is a new and/or infrequently employed drug.

Printed in Switzerland on acid-free paper by
Thür AG Offsetdruck, Pratteln
ISBN 3-8055-6258-6

Contents

Preface

Giuseppe Saggese

There are no doubts that from the first studies of Raben on the use of human GH in GH deficiency, much progress has been made; however, what is impressive is the rate at which this progress has proceeded in the last few years. The most important factor in priming the research in the field of short stature has surely been the advent of recombinant human GH (rhGH), in other words the possibility for the investigator to have unlimited amounts of hGH to treat either short-statured children not necessarily affected by GH deficiency or even patients without short stature but affected by diseases that may benefit from the anabolic action of GH. Obviously, owing to these changes in the therapeutical use of the hormone, some problems and questions arised. One aspect regards the safety and the metabolic consequences of rhGH therapy, considering that at present the practice of giving supraphysiological doses of rhGH is more and more frequent, particularly in non-GH-deficient children. As the hormone is administered in non-GH-deficient children, a crucial point is to assess the real benefit of so expensive a therapy that compels the child to daily injections. Indeed, a practical problem concerns the compliance of the hormone administration and on this matter new devices have been set up and tried. Thus, the underlying concept of this special issue of *Hormone Research* is to give the community of pediatric endocrinologists a current review on advanced research aspects and issues of relevance in the clinical management of children with disturbances of growth and development.

We are confident that these contributions will be stimulating for the pediatric endocrinologist, who is involved in the clinical management of short children daily.

Horm Res 1995;44(suppl 3):2–7

Assunta Albanese[a]
Christine Hall[b]
Richard Stanhope[a]

[a] The Institute of Child Health,
[b] Department of Radiology,
Great Ormond Street Hospital for Children, London, UK

The Use of a Computerized Method of Bone Age Assessment in Clinical Practice

Key Words
Computerized bone age assessment

Abstract
We assessed the reliability and repeatability of a new computerized bone age system, both versions 3.4 and 3.5 (licensed by Discerning System Inc. and Ares Service SA, Serono), able to automatically assess bone age on a left hand and wrist radiograph. This computer system is based upon Tanner and Whitehouse's method (TW2), but there are important differences. Our sample included an initial group of 40 patients who had growth delay/constitutional delay of growth and puberty (n = 10), growth hormone insufficiency/deficiency (n = 15), low birth weight/Silver-Russell syndrome (n = 9), precocious puberty (n = 6), as well as 20 patients with various skeletal dysplasias (multiple epiphyseal dysplasia n = 7, pseudoachondroplasia n = 7, acrodysostosis n = 5, achondroplasia n = 1), 7 girls with Turner syndrome, and 10 boys with nephrotic syndrome on chronic corticosteroid treatment. Multiple anthropometric readings of the same radiographs demonstrated excellent repeatability of the assessment. In addition, the number of times that a manual insertion of a grade was required was similar in four different assessments. The computerized method did not entirely avoid errors in interpretation as the position of the x-ray on the screen was critical. There was a high manual insertion rate in radiographs of children with skeletal dysplasia. However, the computer assessment system, version 3.5, performed adequately for radiographs of children with normal bone morphology and Turner syndrome and had the advantage of a continuous scale of assessment.

Introduction

Measurement of skeletal maturity, expressed as bone age (BA), is a useful tool in assessing a child's development and is independent of chronological age, size, and height. During a child's growth, his/her epiphyseal centres go through definable stages which can be radiographically demonstrated. BA can be assessed from the time of the appearance, and from the stage of maturity, of the different epiphyseal centres. The predominant use of BA is to predict final height. Radiographs of the hand and wrist have been the commonest radiographs used to assess BA. Several methods have been described: the Greulich and Pyle system [1], the Tanner and Whitehouse method (TW2) [2], and recently the Fels method [3, 4]. There is no unanimous agreement on which is optimal and some are

Dr. R. Stanhope
Medical Unit
The Institute of Child Health
30 Guilford Street
London WC1N 1EH (UK)

0301–0163/95/0449–0002
$8.00/0

better in specific diseases states [5]. Indeed, maturation of hand epiphyses are dependent on the endocrine status of the patient: for instance, in the presence of precocious puberty the epiphyses tend to fuse proximally, while in delayed puberty distally. This is one of the reasons why an atlas method such as Greulich and Pyle, where an excessive weight is given to the carpus, has enormous disadvantages [5]. In our practice we use the TW2 method, which gives a series of standard appearances or stages through which radius, ulna, metacarpals, proximal, middle and distal phalanges of the first, third and fifth digits (radius, ulna and short bones, RUS) progress. Each bone is matched with the standard and is then attributed a score which, summed, gives a bone maturity score. The size of the epiphysis related to the metaphysis plays a major role in the assessment of BA, so that any pathological condition which interferes with the development of the cartilaginous growth plate, such as chronic corticosteroid treatment, will produce aberrant results. However, it is important to appreciate that BA assessment is relatively subjective and errors in interpretation occur. In the estimation of final height, although measurement of height is accurate (±0.2%), reliability of the BA rating, even with an experienced assessor, is only within ±0.5 years in 95% of cases using the TW2 method [2]. In addition, because the scoring is discontinuous, a change of one stage in the scoring of a single bone may result in a jump in BA score of >0.3 year. In order to improve height prediction it is necessary to improve bone age ratings and for this reason a computerized analysis system has been developed [6–8].

The aim of our study was to assess the reliability and repeatability of this image analysis system for assessing BA in children with growth and endocrine disorders.

Patients and Methods

The sample included x-rays from a total of 77 patients (33 F, 44 M) who were divided into 2 groups: In the first group (group 1) we included radiographs from 40 patients (mean chronological age 8.9 years, range 2.4–14.8) who had growth delay/constitutional delay of growth and puberty (n = 10), growth hormone insufficiency/deficiency (n = 15), low birth weight/Silver-Russell syndrome (n = 9), precocious puberty (n = 6). In the second group (group 2), radiographs of 20 patients (mean chronological age 9.24 years, range 0.5–20) with skeletal dysplasias (multiple epiphyseal dysplasia n = 7, pseudo-achondroplasia n = 7, acrodysostosis n = 5, achondroplasia n = 1), 7 girls with Turner syndrome (mean 7.4 years, range 4.2–13.6) and 10 boys with nephrotic syndrome on chronic corticosteroid treatment (mean 16.5 years, range 14.4–19.8), were analysed. All the radiographs of the first group were initially rated manually by an experienced assessor (R.S.) in routine clinical practice when the patients attended the growth clinic and afterwards by a novice rater (A.A.) without any knowledge of the first assessment. Before the rating of the radiographs for the present study, A.A. familiarized herself with the TW2 method by assessing a random sample of 100 radiographs under R.S.'s supervision and with the computer-aided estimation of skeletal age by rating 30 radiographs. A.A. was also the computer operator for this study. Each radiograph of the first group was initially assessed using the version 3.4 of the 'Computer-Aided Estimation of Skeletal Age' 4 times in random order, or which 3 occasions used the long method, which analyzes the 13 bones of the standard RUS method, and once by the short method which analyzes 6 bones (radius, ulna and metacarpus, proximal, middle and distal phalanges of the third digit). Thirty-nine out of 40 radiographs from group 1 were reassessed twice by the long method using an updated version 3.5 of this software. During the first occasion (A) the operator entered the grade manually when the difference between the operator and the computer rating was >±0.75 (see below) while during the second occasion (B) the operator manually inserted the grade only when the computer was unable to rate the epiphysis automatically. The radiographs of the second group were assessed twice only using the version 3.4: first they were manually assessed by R.S. and then computer-aided, using the long method, by A.A. who used R.S.'s rating for the initial data selection. The manual method was only used by the experienced assessor in the second group as the novice rater had difficulty in manual assessment relating to the abnormal morphology.

We used versions 3.4 and 3.5 (the up-to-date version 3.5 was only available when the study with version 3.4 was completed) of the Computer-Aided Estimation of Skeletal Age, licensed by Discerning Systems Inc. and Ares Service SA (Serono), 1992-93 (Burnaby, B.C., Canada). The required equipment includes a viewstand on which the radiographs are placed, a camera and lens mounted on the viewstand and a video acquisition and display board, which takes the signal from the video camera and displays the image on the computer screen and a personal computer. In figure 1 a close-up image on the computer monitor is shown during a BA assessment. The software characteristics were reported by Tanner et al. [6, 7]. In order to determine the BA of a radiograph the operator first selects the template (stage A to I of the TW2 method) displayed on the screen which coincides with the operator's assumed rating. After adjusting the focus and zoom as required the operator selects 'Grade' to have the system automatically determine the grade. If the grade seems not to be correct it is possible to select a different template or reposition the bone and repeat. The computer system also offers a continuous scale for the score so that the stepped increase obtained manually is avoided. The computer will display an alert message if there is a major difference between the grade and the template chosen by the operator. The operator can enter the grade manually if the system is unable to grade the bone correctly. The differences between the operator using the manual method and the computer rating were quantified and given a score of 0 when they coincided, ± 1 when there was a ± 1 stage difference and so forth. When the difference between the operator and the computer rating was >±0.75, the operator inserted the grade manually. At least 5 attempts at grading were made when the computer system differed unacceptably from the manual rating. When A.A.'s manual gradings were compared to those of R.S.'s, a 0 score was given when they coincided, ± 1 when A.A.'s manual rating was different in ± 1 stage from those of R.S.'s, and so on. Statistical methods were by one-way analysis of variance (ANOVA) and χ^2 test.

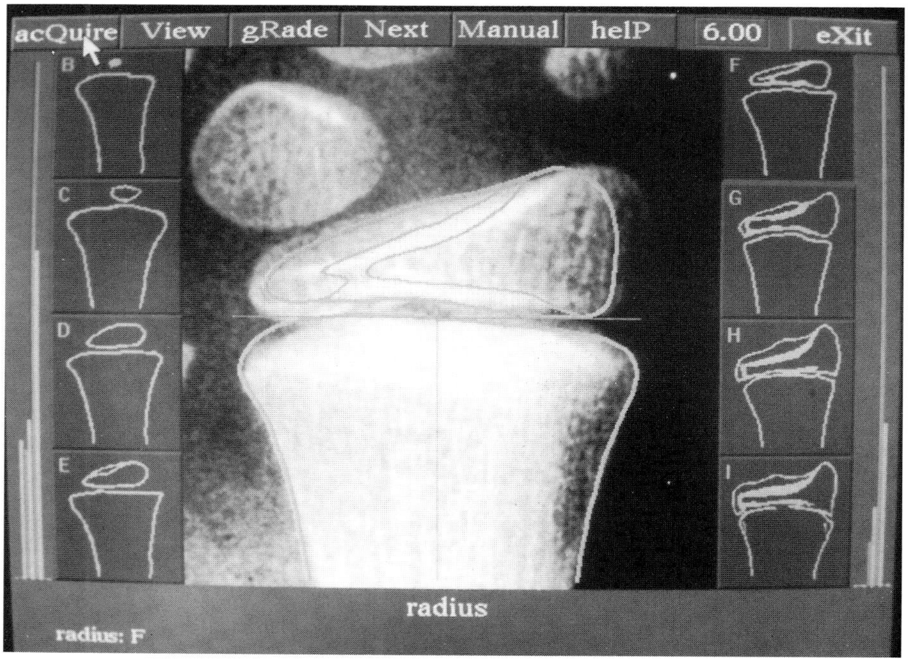

Fig. 1. Computer screen displaying the automatic grading of the radius. The magnified image is on the centre of the screen with computer template surrounding the circumference of the epiphysis and metaphysis. Choice of grading, B to I, are shown on the left- and right-hand side of the screen. The computer letter grade of F is shown at the bottom with a numeric grade of 6.00 at the top right-hand corner.

Results

There were no significant differences among the 6 different ratings in the analysis of the radiographs from group 1 (R.S.'s and A.A.'s manual readings, the 3 different long-method computer-assessed readings and the short-method computer-assessed readings) using version 3.4 of the computerized analysis system. Mean BA readings were, respectively, 7.7, 7.9, 8.1, 8.0, 7.9, and 8.2 'years'. Consequently the reliability and the repeatability of the image analysis system appears to be high. However, the operator had to manually insert the grade in 518 cases out of 1,800 (29%) total requests for an automatic grading. Figure 2a illustrates the degree of difference between the manual and the computer grades for each single epiphysis. The range of percentage of difference was between 6.3% for the third proximal phalanx and 47% for the first metacarpal. Each single epiphysis has a different weight score, with the radius and ulna being the highest in determining the overall bone maturation score. For this reason the high percentage of computer difference for radius and ulna (of 21.9 and 26.9% respectively) implies that without manual insertion, overall computer BA assessment may be unreliable. By contrast, figure 2b shows the percentage of divergence between R.S.'s and A.A.'s manual gradings for each epiphysis. The range of divergence was between 0% for the first metacarpal and 17.5% for the fifth metacarpal, first and third proximal phalanges. The relative divergence between the operator and the computer ratings are demonstrated in figure 3a (number of manual insertions/total requests for an automatic grading). Six different groups of grade were identified (<–1.5, –1.5, –0.75, +0.75, +1.5 and >+1.5). The distribution of the divergence between the two manual assessors are illustrated in figure 3b. A divergence of greater than one stage never occurred manually, in contrast to the computer assessment. We also verified the computer divergence consistency, comparing the total number of manual insertions overall, and the number of manual insertions for each individual bone among the 4 repeated ratings. No significant difference (χ^2 test) was found among the number of times the operator had to insert the grade manually during the 4 repeated readings, and this was the case for both the overall ratings and for each individual bone.

Using version 3.5 there was no significant difference among the overall 4 readings with mean BA values of 7.8 (R.S.), 7.9 (A.A.), 8.0 (occasion A) and 8.2 (occasion B) years. However, the intraindividual range of difference was –2.7 to 1.4 years between manual (A.A.) and automatic (B) ratings, –1.4 to 2.6 years between manual (R.S.) and automatic (B) and –0.8 to 1.4 years between the two manual (A.A. and R.S.) ratings. The automatic rating (B) is plotted versus the manual A.A. and R.S. ratings respectively in figure 4a and b, while figure 4c represents the scatterplot between the two manual ratings (A.A. and R.S.). Manual insertion of the grade was required 207 out

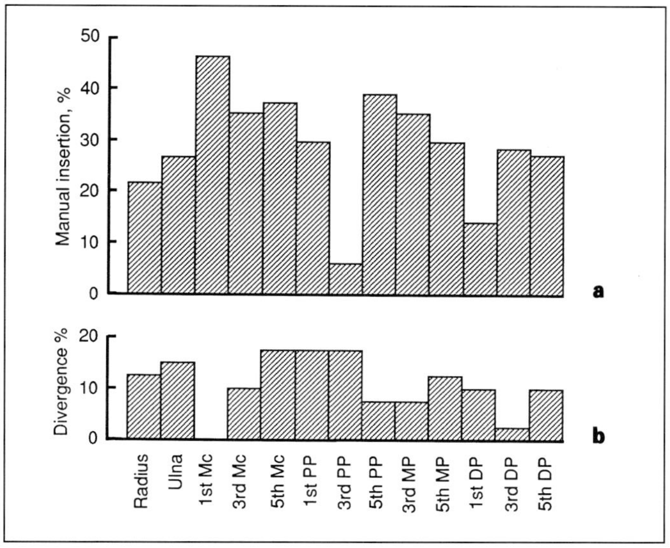

2

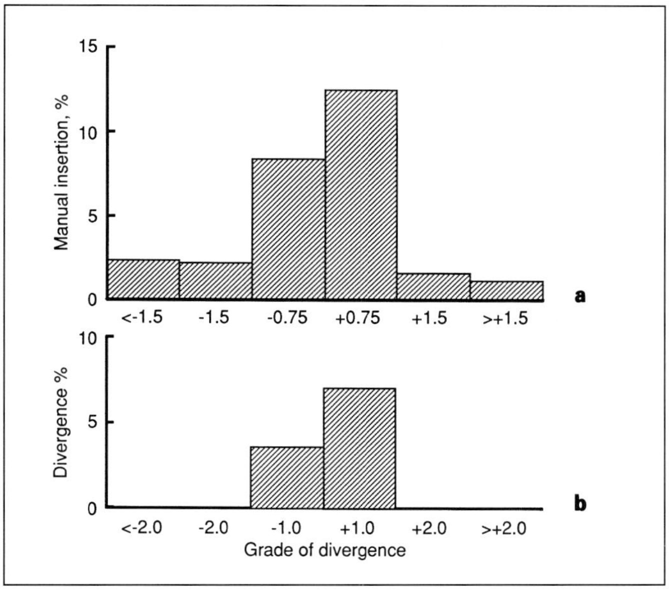

3

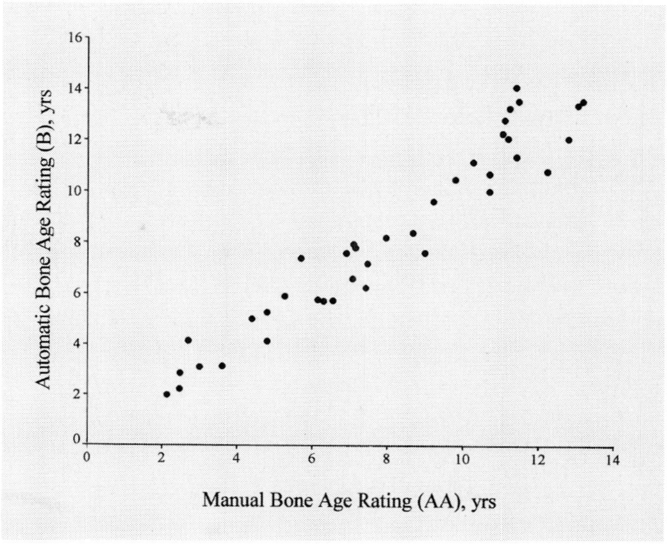

4a

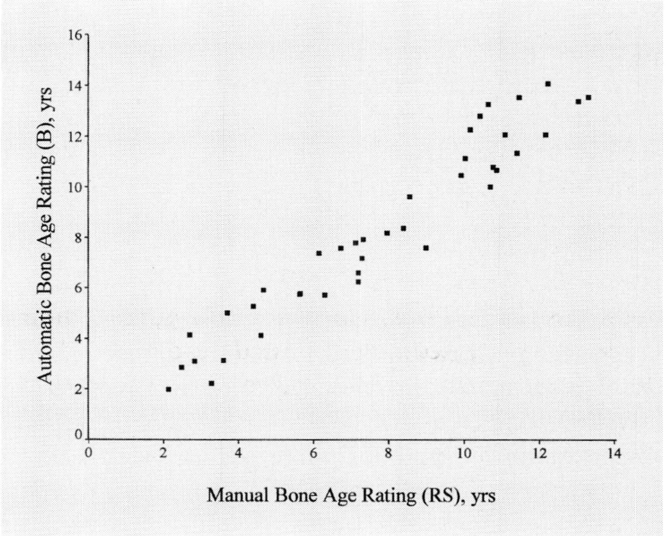

4b

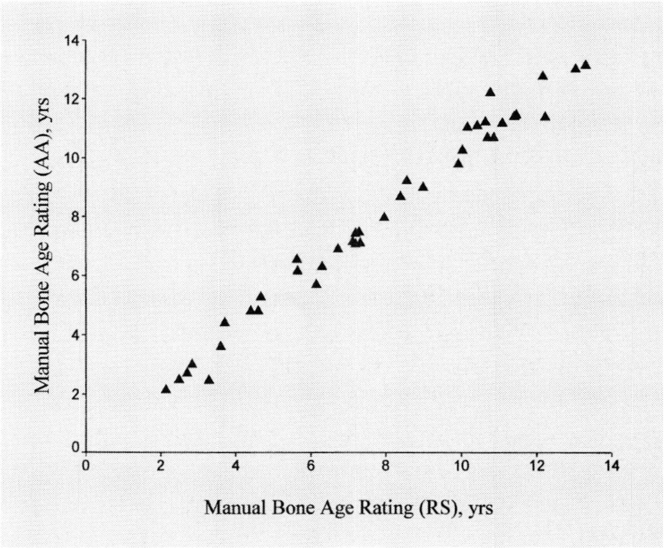

4c

Fig. 2. a Manual insertion rate in the grading of 40 x-rays for each individual epiphysis. The operator inserted manually the grade when the difference between the manual and the computer ratings were $>\pm 0.75$. Mc = Metacarpal; PP = proximal phalange; MP = middle phalange; DP = distal phalange; n = number of epiphyses rated. **b** Divergence between expert and novice assessors's manual gradings of 40 x-rays for each individual epiphysis. Abbreviations as in figure 3a.

Fig. 3. a Rate of manual insertions as a percentage to rate BA in 40 radiographs, each of them assessed 4 times. Six different grades of divergence between manual and computer grading were identified. **b** The percentage distribution of the divergence between the two assessors's manual grading of the same 40 x-rays is reported. Six different grades of divergence were also considered.

Fig. 4. Computerized BA rating (B) plotted against manual rating, by a novice rater (A.A.) (**a**); computerized BA rating (B) plotted against manual rating, by an experienced assessor (R.S.) (**b**); manual BA rating by a novice rater (A.A.) plotted against manual rating by an experienced assessor (R.S.) (**c**).

of 507 (40.8%) total requests during the first automatic rating (A) and only 9 out of 507 during the second (B).

During the computer analysis of the radiographs of patients with skeletal dysplasia (from group 2) the operator had to manually insert the grade 164 times out of 260 (63%). Consequently, no further analysis was carried out in this group. Similar results were obtained for boys with nephrotic syndrome on chronic corticosteroid treatment (48%) in contrast to those of girls with Turner syndrome, where the manual insertion rate (33%) was similar to radiographs with normal morphology from children in group 1.

The process of computer rating using 13 bones score lasted about 30 min per radiograph at the beginning of this study and decreased to 10–15 min after practice. There was no change in the duration of the process of computer rating using version 3.5, compared to version 3.4. Using only a 6-bone assessment the time for rating was reduced to 6–10 min. By contrast, manual assessment by the experienced assessor was always achieved in <2 min, for both rating and addition of the individual bone scores.

Discussion

We initially used this computer-aided method of BA assessment as a computerized TW2 method [6]. However, our findings have led as to believe this hypothesis to be inaccurate. Different criteria are applied to both manual and computerized methods and these are based on different aspects of bone morphology and are not directly comparable. The computerized methods of BA that we have assessed is programmed on normal morphology [6–8], as is the TW2 manual method [2], with no consideration to pathological bone maturation. It has been programmed using hand x-rays of normal children and as such cannot be compared to the interpretation of the manual TW2 method in pathological epiphyseal/metaphyseal morphology by an experienced clinician. However, it was a perfectly acceptable method in radiographs from girls with Turner syndrome, as other authors have demonstrated [9] using the same computerized system.

There is no method for BA assessment which is uniformly acceptable in clinical practice. One of the main problems with all the available methods is related to the low inter-observer reliability. For this reason the availability of a computer system able to automatically estimate BA has been appreciated to be a significant methodological advance. The aim of this study was to assess the reliability and repeatability of this image analysis system. Certainly, the accuracy between the two manual assessors was well within that specified by Tanner et al. [2] and both could be favourably compared to the computer, especially when manually assisted. Our experience suggests, however, that computer assessment of BA obtained with such software is still operator-dependent and requires a knowledge of the TW2 method because at the beginning of each epiphyseal assessment he/she has to choose the appropriate template. In addition, the operator also needs to interpret the grade given by the computer and decide whether to accept it, repeat, or to eventually reject it and insert the grade manually.

There were several reasons why manual insertion of the grade was necessary. When the morphology or the position of the epiphysis/metaphysis was atypical, the reliability of the computer system was inadequate, as illustrated by the 63% manual insertion rate in radiographs from children with skeletal dysplasia, this was especially so in 'coned' epiphyses. Occasionally the sensitivity of the computer to recognize particular stages through which the epiphysis matures was also inadequate: for instance, the presence or absence of capping of the epiphysis, or the relative size of the epiphysis in comparison to that of the metaphysis, or the difference between an atypical fused epiphysis and an overlap between the metaphysis and the epiphysis. When osteoporosis was present or the contrast of the x-ray film was poor, the image obtained on the computer screen was sometimes inadequate and resulted in a subsequent failed computer rating.

The repeatability was high: the computer grading was almost identical when 4 separate ratings were performed on the same radiographs, even though this was mainly due to the many attempts at grading which were carried out. Indeed, there was a variability among the single attempts to rate the same epiphysis according to the position or focus of the film. The number of times the operator had to insert the grade manually was also similar during the 4 ratings with the version 3.4. The number of manual insertions was acceptable for two epiphyses (the third proximal and the first distal phalanges) but not for the other epiphyses and in particular not for the radius and ulna. These two latter bones have the heaviest weighting on total maturity scores so that a mistake in their grades can seriously modify BA assessment. A possible explanation of the improved results from the third proximal and the first distal phalanges grading may be related in the former to its relative constant morphology, and in the latter to its position which allows a good contrasted image on the screen without interference from soft tissue and other bone structures. In addition, the operator developed a skill on how to focus this last epiphysis thereby obtaining a low percentage of manual insertions. In pathological conditions when the structure/size of the epiphysis and metaphysis is abnormal, such as in skeletal dysplasia or chronic corticosteroid treatment, the use of the computerized system was unhelpful. The presence of a warning message on the screen was not always helpful as sometimes, despite the alert message, a 'correct' grade was given. The computer was only rarely incapable of rating an epiphysis.

When the short 6-bone method was used there is an extrapolation of the value of the bones that were not assessed from the values of those tested. For instance, if the grade of the third metacarpal was 'E' the computer system would also grade the first and the fifth metacarpals 'E'. This approach may be reasonable for a normal child but in children with endocrine or growth disorders it may under- or overestimate BA because maturation does not always proceed at the same rate in different bones.

The time interval that the computerized process required was excessive. It takes only 1–2 min for an experienced assessor to rate a radiograph manually. By contrast the computerized assessment was at least 10 min and it is unlikely that this could be reduced further. Sometimes, due to the poor radiograph quality, the operator experienced difficulty in rating it directly on the screen, so that the time required became longer. The equipment was not portable.

The TW2 [2] method has a discontinuous scale of scoring. This implies that a change of rating results in a jump in the BA score and, in addition, when the stage of maturation of a bone is borderline the choice may be extremely subjective. This computer system offers a continuous scale of staging, which theoretically improves the rating but also means that, although based on TW2, this is fundamentally different from TW2. Indeed, '+' or '–' can follow the grade selected so that for instance D+ or E– may be the grade selected when the stage of development of a bone is somewhere between D and E. Until the criteria to define a stage as '+' or '–' has been clarified, an absolute comparison with manual rating is not possible. We have emphasized this by the different grades of divergence in figure 3a (computer) and figure 3b (manual).

In summary, we consider the computerized method, version 3.5, to be satifactory for the assessment of radiographs from normal children and Turner syndrome. However, a skilled operator is a requirement. As for methods using manual assessment, difficulty was experienced in the interpretation of 'bone age' when major morphological abnormalities were present such as in skeletal dysplasias. Although manual methods may be obsolescent, they are not obsolete [7]. The major advantages of the computer-aided method are consistency and a continuous scale of assessment.

Acknowledgements

We are grateful to the Child Growth Foundation for financial support of this study and to Serono UK for the loan of the bone ageing computer. We are grateful to R. Gibbons and D. Patterson, from Discerning Systems Inc., for helpful criticism of our data interpretation.

References

1 Greulich WW, Pyle SI: Radiographic Atlas of Skeletal Development of Hand and Wrist, ed 2. Stanford, Stanford University Press, 1959.

2 Tanner JM, Whitehouse RH, Cameron N, Marshall WA, Healy MJR, Goldstein H: Assessment of Skeletal Maturity and Prediction of Adult Height, ed 2. London, Academic Press, 1983.

3 Roche AF, Roberts J, Hamill PVV: Skeletal maturity of children 7–11 years, United States. National Center for Health Statistics, Series 11, No 140. DHEW Publ No (HRA) 75-1622. Rockville, DHEW, 1974.

4 Roche AF, Roberts J, Hamill PVV:Skeletal maturity of youths 12–17 years, United States. National Center for Health Statistics, Series 11, No 160. DHEW Publ No (HRA) 77-1642. Rockville, DHEW, 1976.

5 Preece MA: Prediction of adult height: Methods and problems. Acta Paediatr Scand 1988; 347(suppl):4–11.

6 Tanner JM, Gibbons RD, Bock D: An image analysis system for TW skeletal maturity. Hormone Res 1992;37(suppl 4):11.

7 Tanner JM, Gibbons RD, Bock D: Advantages of the computer-aided image analysis system for estimating TW skeletal maturity: Increased reliability and a continuous scale. Pediatr Res 1993;33(suppl):S35.

8 Tanner JM, Gibbons RD: Automatic bone age measurement using computerized image analysis. J Pediatr Endocrinol 1994;7:141–145.

9 Van Teunenbroek A, Roks A, De Muinck Keizer-Schrama S, Mulder P, Rongen-Westerlaken C, Drop S: Bone maturation determined by a computerized system in girls with Turner syndrome with and without estrogen therapy. Hormone Res 1994;41:86.

Horm Res 1995;44(suppl 3):8–14

S. Zucchini[a]
B. di Natale[b]
P. Ambrosetto[c]
R. De Angelis[b]
E. Cacciari[a]
G. Chiumello[b]

[a] First Pediatric Clinic, University of Bologna,
[b] Department of Pediatrics, Endocrine Unit, Scientific Institute H.S. Raffaele, University of Milan,
[c] Department of Neuroradiology, Institute of Neurology, University of Bologna, Italy

Role of Magnetic Resonance Imaging in Hypothalamic-Pituitary Disorders

Key Words
Magnetic resonance imaging
Pituitary dwarfism
Diabetes insipidus
Puberty

Abstract
Improvement of MRI diagnostic accuracy in the study of the hypothalamic-pituitary region provides precise anatomic details. In pituitary dwarfism, MRI reveals severe sella/pituitary gland and stalk hypoplasia with or without posterior pituitary ectopia, and empty sella, and this more frequently in patients with multiple pituitary hormone deficiency. Two main hypotheses have been proposed to explain these findings: traumatic stalk transection during breech delivery, and abnormal embryonic development of the pituitary gland. The association between neuroradiological findings and type/severity of endocrine alteration has not yet been clarified. In diabetes insipidus, MRI findings are normal picture, posterior lobe not visible, and thickened stalk (as expression of preclinical/initial histocytosis). Patients with central precocious puberty or hypogonadotropic hypogonadism rarely show morphologic abnormalities (hamartoma of the tuber cinereum, partially empty sella). So far, MRI permits one to identify morphologic pictures in diseases previously considered 'idiopathic'.

Introduction

CT and more recently magnetic resonance imaging (MRI) have enabled the noninvasive study of the hypothalamo-pituitary area in children with endocrine disorders. The former has also identified the picture of 'empty sella' (presence of CSF within the sella turcica replacing the image of the pituitary), above all in children with pituitary deficiency, but also in subjects with disorders of puberty, diabetes insipidus and hyperprolactinemia [1–5]. The investigation by Stanhope et al. [3], who used high-resolution CT, demonstrated the presence of pituitary hypoplasia or aplasia in a high number of subjects with hypopituitarism.

In the last few years, MRI has dramatically improved the diagnostic accuracy of the study of the hypothalamic-pituitary region. If used with the appropriate sections (sagittal and coronal planes of slices $\leqslant 3$ mm thick with T_1-weighted images), MRI provides precise details on the pituitary anatomy (stalk, pituitary dimensions or differentiation between anterior and posterior lobe). The various hypothalamic-pituitary disorders are considered separately.

Prof. Giuseppe Chiumello
Department of Pediatrics
Scientific Institute H.S. Raffaele
Via Olgettina 60
I-20132 Milan (Italy)

0301–0163/95/0449–0008
$8.00/0

Pituitary Dwarfism

The morphological and structural abnormalities of the pituitary revealed so far by MRI can be summarized as follows: severe hypoplasia of both sella turcica and pituitary gland with absence of normal T_1-weighted posterior pituitary hyperintensity; absence or severe hypoplasia of the stalk; nodular hyperintensity at the level of the infundibular region (posterior pituitary ectopia), and presence of intrasellar CSF (empty sella).

The first data describing the sellar structural abnormalities in few cases were published in 1987 [6] and were later confirmed in a larger series of subjects [7–17]. All these studies agreed on the finding of more frequent abnormalities in subjects with multiple pituitary hormone deficiency who are characterized by stalk abnormalities (not visible) possibly with associated ectopic neurohypophysis (fig. 1). Pellini et al. [11] described an invisible stalk in all of their subjects (n = 12) with multiple deficiency and in 8 of 18 (44.4%) subjects with isolated GH deficiency and, similarly, an investigation by the same group [12] found the above abnormalities in 20 of 21 subjects with multiple deficiency (95%) and in 14 of 36 (39%) subjects with isolated GH deficiency. In the latter investigation, which also examined 15 nondeficient subjects as controls, reduced pituitary and sellar volumes were found in the hypopituitaric subjects, regardless of a multiple or isolated hormone deficiency. Patients with disconnected stalk showed sellar and pituitary volumes lower than patients with normal stalk. In the study by Cacciari et al. [13], in which a larger number of subjects including 70 hypopituitary patients was examined, typical intrasellar abnormalities were present in all subjects with multiple pituitary deficiency and only in 12% of subjects with isolated GH deficiency. Here, too, subjects with multiple deficiency showed the highest degree of pituitary hypoplasia. The more recent study by Maghnie et al. [14] provided further information, stating that multiple hormone deficiency is dependent on the type of hypothalamo-pituitary structural abnormality and not on the degree of pituitary hypoplasia. A common agreement of these studies is strict endocrinologic surveillance in the patients with isolated GH deficiency showing nonvisible stalk and posterior pituitary ectopia, since these patients may eventually develop other pituitary hormone deficiencies. Very often subjects with disconnected stalk display ectopia of the posterior lobe at the level of the hypothalamic median eminence, while in the other cases the typical bright signal is not visible, despite the absence of symptoms of diabetes insipidus. An easy explanation of this phenomenon can be

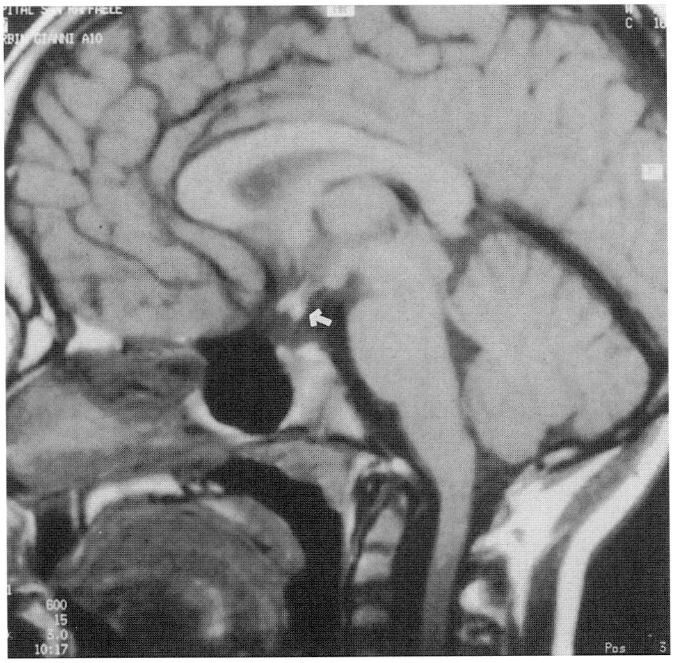

Fig. 1. Sagittal image of the ectopia of the posterior pituitary gland seen as a hyperintense spot (arrow), hypoplastic anterior pituitary gland, absence of the stalk.

found in the study by Brooks et al. [18], showing that the posterior lobe may or may not be visible due to functional reasons.

As regards the presence of intrasellar CSF, the so called 'empty sella', the advent of MRI has dramatically reduced the frequency of such anomaly described until then with CT scan. The investigation by di Natale et al. [19], which, as early as 1987, described with MRI 4 pituitary dwarfs with nonvisible stalk, examined 16 GH-deficient subjects presenting empty sella at CT scan, by comparing the results of the two techniques: the more precise MRI visualization of the hypothalamo-pituitary structures denied the presence in most cases of a pathologically increased intrasellar CSF and identified anterior lobe hypoplasia and thin or absent stalk. The authors hypothesized that the empty sella diagnosed with CT scan were only an epiphenomenon of pituitary hypoplasia, and the entity found in children probably has no relationship to that of adults. The increased CSF would merely reflect the need to fill the sella turcica incompletely occupied by a hypoplastic pituitary. A more recent paper by Cacciari et al. [16] reexamined the matter and studied with MRI a large number of children and young adults (339 cases) affected by hypothalamo-pituitary disorders. In the study, an empty sella was specifically considered when, irrespective of intrasellar abnormalities, CSF is found within the sella together with an upper concave and hypoplastic pituitary anterior lobe which is flattened at the sellar bottom. The authors found an empty sella (pituitary height <2 mm) or a partially empty sella (pituitary height >2 mm) in 35% of subjects with multiple pituitary hormone deficiency (almost always with other intrasellar abnormalities) and in 9% of subjects with isolated GH deficiency (without the simultaneous presence of these anomalies). They concluded that an entity thus defined should be maintained when identified by MRI and can be present independently of stalk abnormalities. Marwaha et al. [20], on the other hand, described 9 patients with isolated GH deficiency, empty sella and normal stalk, as further proof that without stalk abnormalities a multiple deficiency is rare, even with extreme pituitary hypoplasia.

Very recently, Maghnie et al. [21], using a new MRI dynamic technique, described delayed visualization of both anterior and posterior pituitary lobes after injection of gadopentetate dimeglumine in a nonhomogeneous group of patients affected by diabetes insipidus and other pituitary hormone deficiencies. Two subjects were affected by histiocytosis (and thick stalk at MRI) and 2 by idiopathic hypopituitarism (1 with thick stalk). Conversely, a normal response was found in subjects with isolated GH deficiency and small pituitary. According to the authors' hypothesis, a new subset of patients has been identified, affected by diabetes insipidus showing pituitary 'vasculopathy', who are at risk for the development of multiple pituitary hormone deficiency.

Etiology

So far, two main hypotheses have been proposed to explain the typical abnormalities found with MRI in hypopituitaric patients: (1) a traumatic transection of the stalk during a breech delivery with the subsequent formation of an ectopic posterior lobe at the proximal stump, (2) abnormal embryonic development of both anterior and posterior lobes.

The first hypothesis is consistent with the fact that the subjects with multiple deficiency (and hence posterior pituitary ectopia) have an unexplained high incidence of abnormal delivery, particularly breech delivery [22]. Fujisawa himself [6] in 1987, describing the 'transected stalk-breech delivery' association, hypothesized the role of a traumatic birth with subsequent transection of the stalk, and the investigations by Kikuchi et al. [7] and Fujita et al. [23] supported this theory. In particular, the latter described 7 hypopituitary patients with transected stalk and traction of the brain and spinal cord during breech delivery.

By contrast, the second theory, currently sustained by many authors [13–17], supports a congenital origin of intrasellar abnormalities which may be part of midline brain abnormalities. Cacciari et al. [13, 16] for example, found in both of their investigations that several patients with a negative history for adverse perinatal events presented the typical abnormalities and conclude that perinatal damage has little importance in causing such anomalies. The recent study by Triulzi et al. [17] on 101 pituitary dwarfs provided new insights into the pathogenesis of pituitary deficiency. The authors found that the subjects with posterior pituitary ectopia differed from normal pituitary patients in having a higher frequency of multiple pituitary hormone deficiency (49 vs. 12%), breech delivery (32 vs. 7%) and associated congenital brain anomalies (fig. 2–4; 12 vs. 7%). Their conclusion is that pituitary hypoplasia, which was particularly severe in the patients with normal or ectopic posterior lobe but hypoplastic stalk, should not be considered as a consequence of breech delivery, but is the result of a congenital defect, due to a defective induction of mediobasal structure of the brain. This would also explain the peculiar association between stalk and pituitary abnormalities with midline brain anomalies, and thus breech delivery may be part of this complex clinical picture.

Relationships between Morphology and Function

The relationship between morphology and pituitary function has long been studied, and the association between neuroradiological findings on one hand and type or severity of endocrine alteration on the other has not yet been clarified. According to several authors' opinions [13–15, 20], there is a certain degree of relationship between morphology and function in multiple pituitary hormone deficiency, that is almost always associated with 'posterior pituitary ectopia and absent or thin stalk', whereas subjects with isolated GH deficiency suffer from that abnormality much more rarely. In partial disagreement are the data of Triulzi et al. [17] who found a posterior pituitary ectopia in 30 of 67 (45%) subjects with isolated GH deficiency.

It is difficult to find a clear association between pituitary dimensions and the function of the gland in toto. In contrast to Cacciari et al. [13] Kikuchi et al. [7] found that hypopituitary patients with TSH deficiency showed a small-sized pituitary and hypothesized a direct relationship between pituitary height and GH secretion. In fact, GH-deficient subjects had smaller pituitaries, while sub-

2 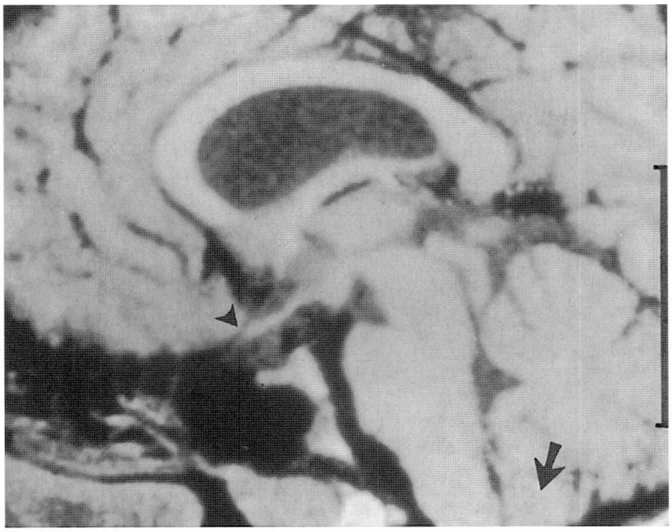

3 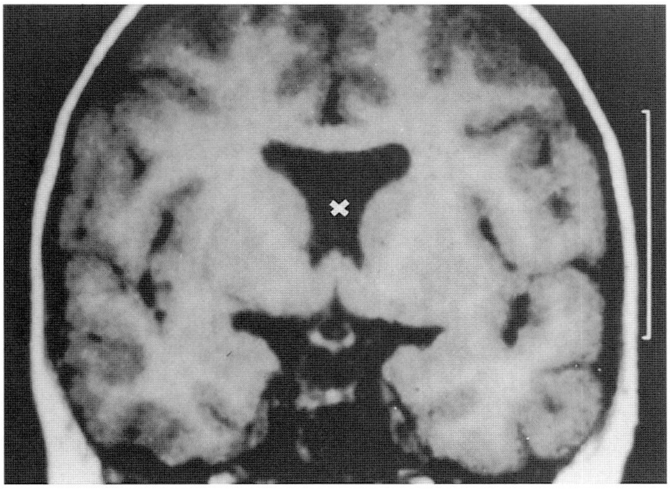

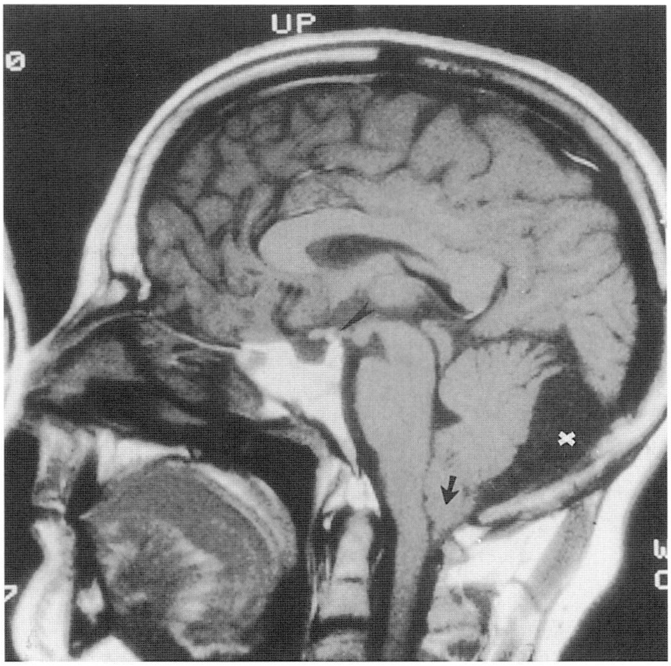

4

Fig. 2. Sagittal image of septo-optic dysplasia: severe hypoplasia of the pituitary gland, the stalk and the chiasma (arrowhead). Partial ectopy of the cerebellar tonsils in the foramen magnum (arrow).

Fig. 3. Coronal image of septo-optic dysplasia: absence of the septum pellucidum evidenced by the asterisk.

Fig. 4. Sagittal image of posterior pituitary ectopia (arrowhead), hypoplastic anterior pituitary, ectopia of the cerebellar tonsils (arrow) and the arachnoid cyst, posterior to the cerebellar hemispheres (asterisk).

jects affected by either hypo- or hyperfunction of gonadotropins showed similar pituitary height values. A partial confirmation of the above hypothesis can be found in the papers by Argyropoulou et al. [15], Kuroiwa et al. [24] and Cicognani et al. [25]. In the first study, in fact, the mostly hypoplastic glands were found in subjects with more severe GH deficiency, and the authors even stated that 'the diagnosis of permanent GH deficiency should be questioned when the pituitary height is normal for age'. In the investigation by Cicognani et al. [25], which differs from the others since the pituitary region of 28 children previously irradiated for acute lymphoblastic leukemia, was examined by MRI, and a positive correlation between GH secretion (arginine test peak and area and mean nocturnal concentration) and pituitary height was found. Furthermore, the patients with pituitary height <4 mm presented a higher percentage of low response to tests. Different conclusions were reached by Kikuchi et al. [7] who found no relationships between GH response to tests and pituitary height, and by Maghnie et al. [14], whose patients' low GH response to GHRH was dependent on stalk abnormality rather than on pituitary hypoplasia.

The abnormalities of the stalk, consistent with suprahypophyseal damage, are in partial agreement with the data obtained with releasing factors GHRH and TRH [26, 27]. In contrast, the investigation of Pellini et al. [11] showed that a pituitary or hypothalamic deficiency was equally distributed between the disconnected and non-disconnected group. However, it should be remembered that the GHRH test can give falsely negative results [28] and cannot be considered a reliable method to identify the location of the damage. Also, Bressani et al. [12] found a poor correlation between MRI results and endocrinological evaluation: 71% of the pituitary dwarfs tested (57 subjects) did not respond to the GHRH test, with similar percentages between the subjects with disconnected (77%) or normal stalk (63%), and no correlations were even found between TSH response to TRH and disconnected stalk. On the other hand, the authors found that an intact stalk was rarely associated with a pathological gonadotropin response to the LHRH test.

To clarify the functional integrity of the hypothalamo-pituitary axis, the study by di Natale et al. [29] investigated the prolactin response to domperidone in 57 pituitary dwarfs affected by isolated or multiple deficiency with or without stalk disconnection and in 29 constitutionally short subjects, used as control subjects. The rationale of the study is consistent with the fact that the dopaminergic functional alteration caused by a transection of the pituitary stalk with damage to the tuberoinfundibular system would cause high basal prolactin levels and a minor response to the antidopaminergic drug. The results were that basal prolactin levels were not different between controls and subjects with stalk disconnection. Domperidone response was lower in pituitary dwarfs than in controls and lower still in subjects with stalk disconnection. The latter, however, showed in some respect a prolactin response, although inferior to that of subjects with anatomical integrity. The authors concluded that the subjects with transected stalk probably maintained a residual anatomical hypothalamic-hypophyseal connection, not detectable by MRI.

Diabetes insipidus

We have already reported that the pituitary posterior lobe is well visible at MRI, appearing as a 'bright spot' both when it is located in the normal position or when it is ectopic at the stalk proximal stump. A normal subject may not display its typical signal probably due to the adiuretine content at that moment [18]. The MRI pictures described in idiopathic diabetes insipidus are the following: normal picture, posterior lobe not visible, thickened stalk.

Fujisawa et al. [30] and others [9] revealed that no subject with diabetes insipidus showed the signal of the posterior lobe; furthermore, 2 out of the 6 subjects with idiopathic diabetes insipidus had a slightly thickened stalk. Besides a subject with idiopathic diabetes insipidus and nonvisible posterior lobe, Chiumello et al. [31] described 1 patient with diabetes insipidus due to a defect in the hypothalamic osmoreceptors in whom neurohypophysis was normally visible. On the other hand, in the study by Cacciari et al. [13], 2 out of the 9 subjects with diabetes insipidus examined showed a normal posterior lobe. The empty sella-diabetes insipidus association, often reported in the literature with CT scan [32, 33], does not seem to be typical of diabetes insipidus when it is not associated with other hormonal deficiencies according to Cacciari et al. [16]. The authors in fact did not disclose an empty sella in any of the 10 subjects with isolated diabetes insipidus. Their conclusion was that the intrasellar CSF per se seemingly had no influence on posterior lobe function.

In the investigation by Maghnie et al. [34], describing 20 patients with diabetes insipidus of various origin, a relationship between low arginine-vasopressin levels and absent posterior pituitary signal was found only in the presence of an organic lesion of the hypothalamic-neurohypophysis axis: in fact, in some of the patients with idiopathic diabetes insipidus, undetectable arginine-vasopressin levels were associated with a normal MRI of the posterior lobe. The same study described 3 patients with diabetes insipidus and thickened stalk out of the 5 examined for already diagnosed histiocytosis suggesting that MRI can be useful to identify preclinical or initial disease with involvement of the hypothalamic-neurohypophyseal tract. Of relevance is the study by Imura et al. [35], in which 9 of 17 patients with idiopathic diabetes insipidus had thickened stalk, enlargement of the posterior lobe or both and at the same time lack of its normal hyperintense signal. The thickening, visible only in subjects with disease duration of less than 2 years, had a tendency to disappear over time, which is a phenomenon already described in histiocytosis [36]. The authors, who even performed two biopsies of the stalk finding a lymphocytic and plasmacellular infiltrate, concluded that their patients had an infundibuloneurohypophysitis on a probable autoimmune basis, which is to be distinguished from the intrasellar localization of Langerhans cell histiocytosis.

Disorders of Puberty

Patients affected by central precocious puberty or hypogonadotropic hypogonadism rarely show morphological abnormalities of the hypothalamic-pituitary region. As regards precocious puberty, the possibility of detecting a hamartoma of the tuber cinereum when the onset of symptoms occurs before the age of 2 years has already been described using CT scan [37]. In the subsequent study of the same group [13], 21 patients with central precocious puberty were examined, and the hamartoma was

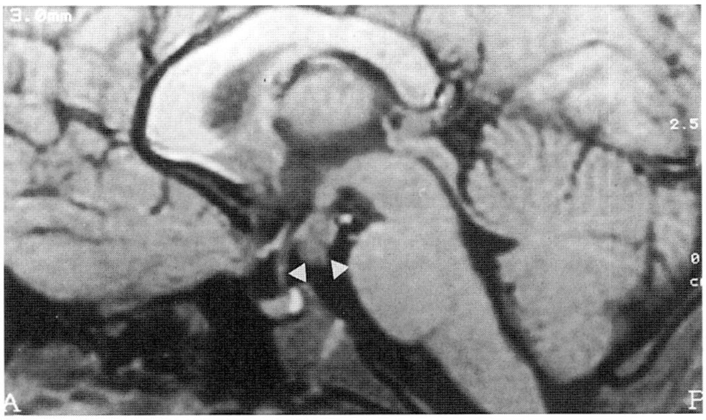

Fig. 5. Sagittal image of a 4.7-year-old girl with precocious puberty and hamartoma at the level of the hypothalamic median eminence (arrowheads). The image of the anterior and posterior lobe is normal.

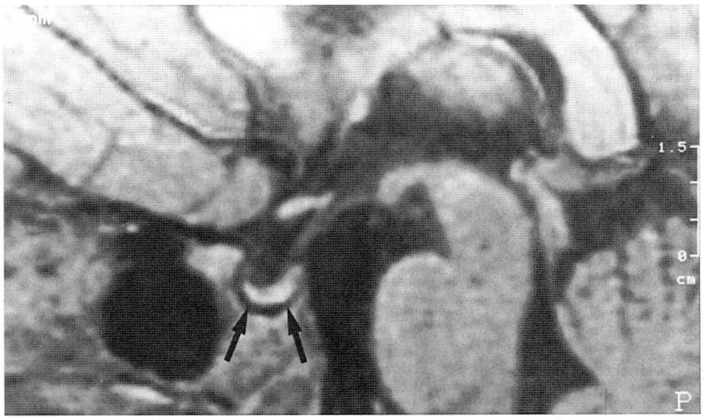

Fig. 6. Sagittal image of a 5.1-year-old girl with precocious puberty and partially empty sella. The hypoplastic and concave pituitary tissue is flattened at the sellar bottom (arrows). The stalk is normally visible.

confirmed in 3 patients (fig. 5; 14%). Six subjects affected by hypogonadotropic hypogonadism were also investigated, and no MRI abnormalities of the sellar area were found. In the following study [16], the authors evaluated 47 patients with precocious puberty and described a partially empty sella in 2 cases (fig. 6; 4.5%). Such an association, apparently exceptional as it combines hypoplasia and hyperfunction of the pituitary, is not surprising, since it had already been described both with CT and MRI [2, 33, 38]. The presence of intrasellar CSF was also identified in 1 of 17 cases affected by hypogonadotropic hypogonadism (5.9%) and in as many as 2 of 5 cases affected by a nonfamilial delay of puberty.

Even if the subjects with empty sella (and hence with severe pituitary hypoplasia) may present either precocious puberty or hypogonadotropic hypogonadism, a certain relationship between pituitary dimensions and gonadotropin secretion seems to exist. First of all, the observation that pituitary height measured with MRI shows an increase from birth to the age of 15 years [39], showing a peak at puberty (when GH secretion increases as well), suggests a probable link between the two parameters. The investigation by Perignon et al. [40] examined the relationship between pituitary dimensions and gonadotropin secretion by studying pituitary heights of 47 girls with breast development which appeared before the age of 8 years. The patients were subdivided into cases with isolated premature thelarche and cases with true precocious puberty: indeed, only the subgroup of the girls with precocious puberty and LH/FSH peak ratio after LHRH test >1 showed pituitary dimensions higher than control values for age. A positive correlation between pituitary height and LH/FSH peak ratio was also found. The authors concluded that the changes in pituitary dimensions apparently reflected the modified activation of the hypothalamic-pituitary-gonadal axis, thus possibly providing an indication of its future development. Similar conclusions have been more recently reached by Sharafuddin et al. [41] who found that a pituitary convex upper margin associated with increased dimensions (height and area) are strongly predictive of central precocious puberty.

Finally, also the preliminary data of Zucchini et al. [42], who compared the pituitary heights of the patients affected by hypogonadotropic (9 cases) or hypergonadotropic (22 cases) hypogonadism, seem to confirm the influence of gonadotropin secretion on pituitary dimensions: the pituitary heights of hypergonadotropic patients, in fact, were significantly higher than those of hypogonadotropic patients.

Conclusions

In recent years, MRI has enabled a noteworthy improvement in the knowledge of the morphology and the etiology of principal hypothalamic-pituitary disorders. As regards etiology, it has been possible to identify specific morphologic pictures in diseases previously considered as 'idiopathic'. Parallel studies with MRI and molecular biology on the genetics of the pituitary hormone activators will probably determine further progress in the understanding of these matters.

References

1 Costigan DC, Daneman D, Harwood-Nash D, et al: The 'empty sella' in childhood. Clin Pediatr 1983;23:437–440.
2 Shulman DI, Martinez CR, Bercu BB, Root AW: Hypothalamic-pituitary dysfunction in primary empty sella syndrome in childhood. J Pediatr 1986;108:540–544.
3 Stanhope R, Hindmarsh P, Kendall B, Brook CGD: High resolution CT scanning of the pituitary gland in growth disorders. Acta Paediatr Scand 1986;75:779–786.
4 Surtees R, Adams J, Price D, et al: Association of adverse perinatal events with an empty sella turcica in children with growth hormone deficiency. Horm Res 1987;28:5–12.
5 Pocecco M, De Campo C, Marinoni S, et al: High frequency of empty sella syndrome in children with growth hormone deficiency. Helv Paediatr Acta 1988;196:92–97.
6 Fujisawa I, Kikuchi K, Nishimura K, et al: Transection of the pituitary stalk: Development of an ectopic posterior lobe assessed with MR imaging. Radiology 1987;165:487–489.
7 Kikuchi K, Fujisawa I, Momoi T, et al: Hypothalamic-pituitary function in growth hormone-deficient patients with pituitary stalk transection. J Clin Endocrinol Metab 1988;67: 817–823.
8 Scotti G, Triulzi F, Chiumello G, di Natale B: New imaging techniques in endocrinology: Magnetic resonance of the pituitary gland and sella turcica. Acta Paediatr Scand Suppl 1989; 356:5–14.
9 Root AW, Martinez CR, Muroff LR: Hypothalamic high-intensity signal identified by magnetic resonance imaging in children with idiopathic anterior hypopituitarism: Evidence suggestive of ectopic posterior pituitary gland. AJDC 1989;143:366–367.
10 Zucchini S, Carlà G, Brayda G, et al: Magnetic resonance imaging (MRI) of the sellar area in 101 patients with different endocrine disorders (abstract). Horm Res 1989;31(suppl 1):23.
11 Pellini C, di Natale B, De Angelis R, et al: Growth hormone deficiency in children: Role of magnetic resonance imaging in assessing aetiopathogenesis and prognosis in idiophatic hypopituitarism. Eur J Pediatr 1990;149:536–541.
12 Bressani N, di Natale B, Pellini C, Triulzi F, Scotti G, Chiumello G: Evidence of morphological and functional abnormalities in the hypothalamus of growth-hormone-deficient children: A combined magnetic resonance imaging and endocrine study. Horm Res 1990;34:189–192.
13 Cacciari E, Zucchini S, Carlà G, et al: Endocrine function and morphological findings in patients with disorders of the hypothalamo-pituitary area: A study with magnetic resonance. Arch Dis Child 1990;65:1199–1202.
14 Maghnie M, Triulzi F, Larizza D, et al: Hypothalamic-pituitary dysfunction in growth hormone-deficient patients with pituitary abnormalities. J Clin Endocrinol Metab 1991;73:79–83.
15 Argyropoulou M, Prignon F, Brauner R, Brunelle F: Magnetic resonance imaging in the diagnosis of growth hormone deficiency. J Pediatr 1992;120:886–891.
16 Cacciari E, Zucchini S, Ambrosetto P, et al: Empty sella in children and adolescents with possible hypothalamic-pituitary disorders. J Clin Endocrinol Metab 1994;78:767–771.
17 Triulzi F, Scotti G, di Natale B, et al: Evidence of a congenital midline brain anomaly in pituitary dwarfs: A magnetic resonance imaging study in 101 patients. Pediatrics 1994;93:409–416.
18 Brooks BS, Gammal TE, Allison JD, Hoffman WH: Frequency and variation of the posterior pituitary bright signal on MR images. AJR 1989;153:1033–1038.
19 di Natale B, Scotti G, Pellini C, et al: Empty sella in children with pituitary dwarfism: Does it exist? Pediatrician 1987;14:246–252.
20 Marwaha R, Menon PSN, Jena A, Pant C, Sethi AK, Sapra ML: Hypothalamo-pituitary axis by magnetic resonance imaging in isolated growth hormone deficiency patients born by normal delivery. J Clin Endocrinol Metab 1992;74:654–659.
21 Maghnie M, Genovese E, Aricò M, et al: Evolving pituitary hormone deficiency is associated with pituitary vasculopathy: Dynamic MR study in children with hypopituitarism, diabetes insipidus and Langerhans cell histiocytosis. Radiology 1994;193:493–499.
22 Van den Broeck J, Vanderschueren-Lodeweyckx M, Malvaux P, et al: Growth hormone deficiency: A hidden obstetrical trauma? Eur J Obstet Gynecol Reprod Biol 1987;26:329–334.
23 Fujita K, Matsuo N, Mori O, et al: The association of hypopituitarism with small pituitary, invisible stalk, type 1 Arnold-Chiari malformation, and syringomyelia in seven patients born in breech position: A further proof of birth injury theory on the pathogenesis of idiopathic hypopituitarism 1992;151:266–270.
24 Kuroiwa T, Okabe Y, Hasuo K, Yasumori K, Mizushima A, Masuda K: MR imaging of pituitary dwarfism. AJNR 1991;12:161–164.
25 Cicognani A, Cacciari E, Carlà G, et al: Magnetic resonance imaging of the pituitary area in children treated for acute lymphoblastic leukemia with low-dose (18-Gy) cranial irradiation. Am J Dis Child 1992;146:1343–1348.
26 Granier PE, Roger M, Chaussain JL, Canlorbe P, Job JC: Prolactin and thyrotropin responses to thyroliberin (TRH) in patients with growth hormone deficiency: A study in 167 patients. Acta Endocrinol (Copenh) 1983;103:433–440.
27 Schriock EA, Lustig RH, Rosenthal SM, Kaplan SC, Grumbach MM: Effect of growth hormone releasing hormone (GHRH) on plasma GH in relation to magnitude and duration of GH deficiency in 26 children and adults with GH isolated deficiency or multiple pituitary hormone deficiency. J Clin Endocrinol Metab 1984;58:1043–1049.
28 Vance ML, Borges JLC, Kaiser DL, et al: Human pancreatic growth hormone-releasing factor: Dose-response relationship in normal man. J Clin Endocrinol Metab 1984;58:838–844.
29 di Natale B, Pellini C, Ackermann S, De Angelis R, Chiumello G: Persisting functional connection in growth-hormone-deficient patients with a transected stalk. Horm Res 1994;41: 193–196.
30 Fujisawa I, Nishimura K, Asato R, et al: Posterior lobe of the pituitary in diabetes insipidus: MR findings. J Comput Assist Tomogr 1987; 11:214–225.
31 Chiumello G, di Natale B, Pellini C, Beneggi A, Scotti G, Triulzi F: Magnetic resonance imaging in diabetes insipidus. Lancet 1989; i:901.
32 Nass R, Engel M, Stoner E, et al: Empty sella syndrome in childhood. Pediatr Neurol 1986;2: 224–229.
33 Scirè G, Cianfarani S, Spadoni GC, et al: Primary empty sella and endocrinopathies in childhood: High prevalence among children with precocious puberty. Eur J Pediatr 1988; 147:665–666.
34 Maghnie M, Villa A, Aricò M, et al: Correlation between magnetic resonance imaging of posterior pituitary and neurohypophyseal function in children with diabetes insipidus. J Clin Endocrinol Metab 1992;74:795–800.
35 Imura H, Nakao K, Shimatsu A, et al: Lymphocytic infundibuloneurohypophysitis as a cause of central diabetes insipidus. N Engl J Med 1993;329:683–689.
36 Tien R, Newton Th, McDermott MW, Dillon WP, Kurcharczyk J: Thickened pituitary stalk on MR images in patients with diabetes insipidus and Langerhans cell histiocytosis. AJNR 1990;11:703–708.
37 Cacciari E, Frejaville E, Cicognani A, et al: How many cases of true precocious puberty in girls are idiopathic? J Pediatr 1983;102:357–360.
38 Rappaport R, Logrono R: Primary empty sella syndrome in childhood: Association with precocious puberty. Clin Pediatr 1991;30:466–471.
39 Elster AD, Chen MYM, Williams DW III, Key LL: Pituitary gland: MR imaging of physiologic hypertrophy in adolescence. Radiology 1990; 174:681–685.
40 Perignon F, Brauner R, Argyropoulou M, Brunelle F: Precocious puberty in girls: Pituitary height as an index of hypothalamo-pituitary activation. J Clin Endocrinol Metab 1992; 75:1170–1172.
41 Sharafuddin MJ, Luisiri A, Garibaldi LR, et al: MR imaging diagnosis of central precocious puberty: Importance of changes in the shape and size of the pituitary gland. Am J Roentgenol 1994;162:1167–1173.
42 Zucchini S, Ambrosetto P, Tani G, et al: Disorders of the hypothalamo-pituitary-gonadal axis and pituitary morphology: A study with MRI (abstract No. 256). Eur Soc Pediatr Endocr 33rd Annu Meet, Maastricht, 1994.

Horm Res 1995;44(suppl 3):15–17

A. Albanese
R. Stanhope

Medical Unit, Institute of Child Health,
London, UK

Growth Hormone, Growth Factors and the Gonad

Key Words
Gonads
Growth hormone
IGF-1
Sex steroids

Abstract
During puberty, the growth spurt and stage of pubertal maturation are two processes intimately related both being dependent on the combined action of growth hormone (GH) and sex steroids. While it is well established that both GH and sex steroids are required for a pubertal growth spurt, the precise role played by GH on the timing of onset and rate of progression through different stages of pubertal maturation is not entirely clear. A better understanding of the involved physiologic mechanisms will help clinicians in the management of adolescents with growth and/or pubertal disorders.

Growth Spurt and Puberty

The onset of the spontaneous growth spurt and pubertal maturation occurs in a certain sequence which is a characteristic of the normal harmony of puberty. Onset of breast development in girls [1] and increase in testicular size in boys [2] are the first signs of puberty and occur at a relatively similar average age, 11.4 and 12.0 years respectively. However, until the onset of puberty there is little difference between the sexes in the pattern of growth. During puberty a different pattern of growth is exhibited. The growth spurt occurs early with the onset of breast development in girls, while in boys towards the end of puberty [1, 2]. Indeed, boys continue to growth at a prepubertal rate until they attain a genitalia stage 3–4 (equivalent to a 10-ml testicular volume) when their growth commences to accelerate. This sequence between sexual maturation and growth is also maintained when the timing of onset of puberty is altered, such as in precocious or delayed puberty. A loss of this harmony requires further investigations as it points to endocrine dysfunction.

Mechanism of the Growth Spurt

Growth hormone (GH) is secreted in an intermittent pulsatile manner, predominantly at night, in a frequency of approximately one pulse every 3 h [3]. During puberty, mean GH [4] and IGF-1 [5] concentrations increase as boys and girls enter the rapid growth phase and decrease as final height is attained. The increased circulating GH concentration mainly occurs as a result of an increase in the amplitude of GH secretion rather than in pulse frequency [6]. As in patients with isolated gonadotrophin deficiency no pubertal growth spurt occurs unless sex steroids are administered, gonadal steroid hormones have been thought to play a central role in the pubertal growth spurt. On the other hand, in the absence of normal GH and IGF-1 secretion, sex steroids only cause a growth spurt which is minimal when compared to the normal pubertal growth spurt so that the contemporary presence of sex steroids and growth hormone is required. The growth-promoting action of sex steroids could be partly related to their stimulatory effect on GH secretion. For many years it has been appreciated that exogenous sex steroids can 'prime' the pituitary gland to secrete more GH during pharmacological or physiological tests of GH stimulation. Oestrogens are considered to be the main media-

Dr. R. Stanhope
Medical Unit
Institute of Child Health
30 Guilford Street
London WC1N 1EH (UK)

0301–0163/95/0449–0015
$8.00/0

tors of this phenomenon in both sexes and in boys they can derive from testosterone by aromatization. The main support to this hypothesis is given by a study [7] which showed significant correlation between circulation levels of oestradiol and not testosterone with 24-hour integrated concentrations of GH and GH pulsatility in boys. Similar results where also found in girls. However, other mechanisms for sex hormone growth-promoting action should be presumed as oxandrolone, an anabolic steroid [8], and dihydrotestosterone [9], both nonaromatizable, can induce a growth acceleration in patients with constitutional delay of growth and puberty without altering GH production. In addition, the treatment of growth delay/delayed puberty with either testosterone or oxandrolone in boys with isolated GH deficiency on a constant GH dose regimen can induce a growth spurt despite the constant GH dose regimen [10]. Furthermore, doubling GH dose regimen at the onset of the pubertal growth spurt in patients with isolated GH deficiency did not significantly increase growth rate [11]. In vitro both androgen and oestrogen receptors have been demonstrated on cartilage cells and local administration of testosterone to the epiphyseal plate stimulates longitudinal bone growth [12]. The cartilage cells from children in the early phase of puberty have the best response to sex steroids [13]. From these data it can be hypothesized that the pubertal growth spurt can occur even in the absence of increased GH secretion and that sex steroids can have a direct peripheral effect, such as sensitization of the cartilaginous growth plate to GH or an enhancement of the bioactivity of GH, in addition to those mediated via an increase in GH secretion. However, the latter may be a secondary phenomenon.

Growth Hormone and Pubertal Development

Sex steroids are secreted from the gonad in response to pulsatile gonadotrophin secretion, which is controlled by a hypothalamic gonadotrophin-releasing hormone (GnRH). During childhood, low amplitude pulses of gonadotrophins are secreted so that the gonads are not completely quiescent and small quantities of gonadal steroid hormones are secreted, which, however, do not seem to alter GH secretion at this stage. During late prepuberty and early puberty there is a progressive increase in both frequency and amplitude of gonadotrophin secretion, initially mainly during the night-time and later in puberty throughout day and night.

There is increasing evidence to suggest that there is a relationship between GH and gonadal maturation with GH playing an important role. The onset of puberty is usually delayed in boys with isolated GH deficiency and treatment with GH does not appear to bring forward the onset of puberty which remains delayed [14]. Also, boys with GH receptor dysfunction (Laron syndrome) have delayed onset of pubertal maturation and both sexes do not have a proper pubertal growth spurt, despite the fact they attain full sexual maturation [15]. GH treatment in boys and girls with isolated GH deficiency appears to accelerate the rate of pubertal maturation [11, 14]. Similar results were previously described in the female rhesus monkey [16], where GH administration increased the rate of gonadal maturation without altering the timing of onset of puberty. More recent data [11] suggest that this phenomenon may be dose dependent with patients treated with higher dose regimens progressing through puberty faster, limiting therefore the efficacy of treatment. The interaction between GH and gonad can also be the major limiting factor in the effectiveness of exogenous GH when used to improve final height in short patients not GH deficient/insufficient. The improved short-term growth rate induced by GH treatment in these patients may not result in an improvement of final height due to the reduction of height prognosis during puberty. So far, only in girls with Turner syndrome has GH growth-promoting action [17] resulted in an improvement of final height. Furthermore, GH treatment appears to be more effective in boys with multiple pituitary hormone deficiency compared to those with isolated GH deficiency [18]. As in both Turner syndrome and multiple pituitary hormone deficiency, pubertal maturation is usually induced exogenously, a manipulation of the timing of onset of pubertal maturation, with a combined approach of a GnRH analogue and GH treatment, may be used, especially in the presence of an inadequate height prognosis. However, as delayed onset of puberty [19] may also cause an insufficient spinal growth with the attainment of eunuchoid body proportion, the benefits of such an approach must be carefully evaluated.

Numerous experiments, both in vitro and in vivo, have demonstrated that GH treatment can increase ovarian response to gonadotrophins through a modulation of the local concentration of IGF-1. Indeed, IGF-1 behaves as a co-gonadotrophin, increasing the ovarian response to FSH, enhancing differentiation of the granulosa cells and inducing LH receptors [20, 21]. Human GH has, therefore, been used to improve ovarian response to gonadotrophin therapy. Women with amenorrhoea associated with oestrogen deficiency [22] require lower doses of gonadotrophins and shorter duration of treatment to ovulate when human GH supplementation is given in addition. Also amenorrhoeic patients with hypogonadotrophic hypogonadism [23] seem to benefit from GH administration, while the group of patients undergoing ovarian stimulation for in vitro fertilization who may benefit from GH administration has to be defined. In in vitro experiments, GH also has a direct effect on oestradiol production by human granulosa cells [24]. IGF-1 has also been implicated in the pathogenesis of polycystic ovaries: it has been hypothesized that insulin may indirectly regulate the production of IGF-1 through its suppression of IGFBP-1, which is an inhibitor of IGF-1 bioavailability. Suppressed IGFBP-1 levels [25], induced by increased levels of insulin, may increase IGF-1 concentration and then amplify the LH-induced androgen synthesis.

Conclusion

Although the close relationship between growth and puberty is well known, much remains to be learnt about the mechanisms by which GH can interact with pubertal development and vice versa. Vital to our comprehension in this field will be the identification of the possible functions of GH and IGF-1 in ovarian physiology, and the effect and mechanism of action of hormones and growth factors regulating cartilage growth. Manipulation of the timing of onset of pubertal development, using a GnRH analogue in addition to GH treatment, can theoretically be an approach to improve final height attainment in short patients.

References

1 Marshall WA, Tanner JM: Variations in the pattern of pubertal changes in girls. Arch Dis Child 1969;44:291–303.
2 Marshall WA, Tanner JM: Variations in the pattern of pubertal changes in boys. Arch Dis Child 1970;45:13–23.
3 Brook CGD, Hindmarsh PC, Stanhope R: Growth and growth hormone secretion: A review. J Endocrinol 1987;119:179–184.
4 Zadik Z, Chalew SA, McCarter RJ, Meistas M, Kowarski AA: The influence of age on the 24-hour integrated concentration of growth hormone in normal individuals. J Clin Endocrinol Metab 1985;60:513–516.
5 Blum WF, Albertsson-Wikland K, Rosberg S, Ranke MB: Serum levels of insulin-like growth factor 1 (IGF-1) and IGF binding protein 3 reflect spontaneous growth hormone secretion. J Clin Endocrinol Metab 1993;76:1610–1616.
6 Hindmarsh PC, Matthews DR, Brook CGD: Growth hormone secretion in children determined by time series analysis. Clin Endocrinol 1988;29:33–44.
7 Ho KY, Evans WS, Blizzard RM, et al: Effects of sex and age on the twenty-four-hour profile of growth hormone secretion in man: Importance of endogenous estradiol concentrations. J Clin Endocrinol Metab 1987;64:51–58.
8 Malhotra A, Poon E, Tse WY, Pringle PJ, Hindmarsh PC, Brook CGD: The effect of oxandrolone on the growth hormone and gonadal axes in boys with constitutional delay of growth and puberty. Clin Endocrinol 1993;38: 393–398.
9 Keenan BS, Richards GE, Ponder SW, Dallas JS, Nagamani M, Smith E: Androgen-stimulated pubertal growth: The effects of testosterone and dihydrotestosterone on growth hormone and insulin-like growth factor-1 in the treatment of short stature and delayed puberty. J Clin Endocrinol Metab 1993;76:996–1001.
10 Albanese A, Stanhope R: Treatment of growth delay in boys with isolated growth hormone deficiency. J Eur Endocrinol 1994;130:65–69.
11 Stanhope R, Albanese A, Hindmarsh P, Brook CGD: The effects of growth hormone therapy on spontaneous sexual development. Horm Res 1992;38(suppl 1):9–13.
12 Young IR, Mesiano S, Hintz R: Growth hormone and testosterone can independently stimulate the growth of hypophysectomized prepubertal lambs without any alteration in circulating concentrations of insulin-like growth factors. J Endocrinol 1989;121:563–570.
13 Blanchard O, Tsagris L, Rappaport R, Duval-Beaupere G, Corvol M: Age-dependent responsiveness of rabbit and human cartilage cells to sex steroids in vitro. J Steroid Biochem Mol Biol 1991;40:711–716.
14 Darendeliler F, Hindmarsh PC, Preece MA, Cox L, Brook CGD: Growth hormone increases rate of pubertal maturation. Acta Endocrinol 1990;122:414–416.
15 Laron Z, Lilos P, Klinger B: Growth curves for Laron syndrome. Arch Dis Child 1993;68:768–770.
16 Wilson ME, Gordon TP, Rudman CG, Tanner JM: Effects of growth hormone on the temporosexual maturation in female rhesus monkeys. J Clin Endocrinol Metab 1989;68: 29–38.
17 Rosenfeld RG, Frane J, Attle KM, Brasel JA, Burstel S, Cara JF, et al: Six-year results of a randomized, prospective trial of human growth hormone and oxandrolone in Turner syndrome. J Pediatr 1992;121:49–55.
18 Burns EC, Tanner JM, Preece MA, Cameron N: Final height and pubertal development in 55 children with idiopathic growth hormone deficiency treated for between 2 and 15 years with human growth hormone. Eur J Pediatr 1981;137:155–164.
19 Albanese A, Stanhope R: Does constitutional delayed puberty cause segmental disproportion and short stature? Eur J Pediatr 1993;152:293–296.
20 Adashi EY, Resnick CE, Svobota ME, Van Wyk JJ: Somatomedin C synergizes with FSH in the acquisition of projection of biosynthetic capacity by altered rat granulosa cells. Endocrinology 1985;116:2315–2342.
21 Adashi EY, Resnick CE, Svobota ME, Van Wyk JJ: Somatomedin C enhances induction of luteinizing hormone receptors by follicle-stimulating hormone in cultured nongranulosa cells. Endocrinology 1985;116:2369.
22 Homburg R, West C, Torresani T, Jacobs HS: Cotreatment with human growth hormone and gonadotropins for induction of ovulation: A controlled clinical trial. Fertil Steril 1990;53: 254–260.
23 Burger HG, Kovacs GT, Polson DM, McDonald J, McClloud PI, Harrop M, et al: Ovarian sensitization to gonadotrophins by human growth hormone. Persistence of the effect beyond the treated cycle. Clin Endocrinol 1991;35:119–221.
24 Mason HD, Martinkainen H, Beard RW, Anyaoku V, Franks S: Direct effect of growth hormone on oestradiol production by human granulosa in vitro. J Endocrinol 1990;126:R1–R4.
25 Conover CA, Lee PDK: Insulin regulation of insulin-like growth factor-binding protein production in cultured HepG2 cells. J Clin Endocrinol Metab 1990;70:1062–1067.

Horm Res 1995;44(suppl 3):18–25

J.P. López Siguero
M.J. Martínez Aedo
M.D. López Moreno
A. Martínez Valverde

Pediatric Endocrinology Section,
Deptartment of Pediatrics,
Children's Hospital of Málaga, Spain

Treatment with Growth Hormone

What Do Children Know and How Do They Accept It?

Key Words
Growth hormone, treatment
Knowledge
Acceptance

Abstract
Treatment with growth hormone in children is a very effective promotor of growth. The psychosocial environment of children may influence the results of this treatment. It has been shown that there is a clear relationship between the knowledge and understanding a child has of the treatment and the degree of compliance and acceptance. We have studied 90 children treated with growth hormone who filled in a questionnaire in order to know the knowledge and acceptance of their treatment. The scores for acceptance were significantly higher than for knowledge ($p < 0.00001$), without correlations between the two variables. The higher marks were achieved by children who had attended an educational camp. A serious lack of knowledge about the aspects directly related with the injection system was noted. We advise to design a suitable educational program for these patients.

Introduction

Treatment with growth hormone (GH) in children with widely diverging etiologies involves a series of difficulties, not only in terms of the diagnosis, but also in terms of evaluating the effectiveness of the treatment.

For 1.5 years, we studied the psychosocial environment of a group of children on GH treatment (unpublished data). The conclusions drawn clearly show that the treatment is not well accepted, which, combined with little knowledge of the preparation, application, indication and completion of the treatment, can result not only in poor compliance, but also in changes in family dynamics, with serious consequences.

Generally speaking, these patients do not receive any specific, programed education with respect to their problem. On occasion, they do not even receive basic information about the treatment and what to expect from it. Normally, the information is directed to the parents, in part due to the age of the patients at the start of treatment and also because the parents are the most interested in the treatment, and it is them who seek help.

In a recent study on English children treated with GH [1], a clear relationship has been found between the knowledge and understanding a child has of treatment and his/her degree of compliance and acceptance. In an effort to verify the knowledge and acceptance registered in our patients, we used Smith's questionnaire, with minor modifications to adapt to our social environment.

Material and Methods

All of the patients were older than 8 years and had normal intelligence. They were receiving GH treatment and were periodically reviewed in the Children's Hospital of Málaga. Each one filled in the questionnaire in the presence of a member of the medical team.

J.P. López Siguero
Pediatric Endocrinology Section, Department of Pediatrics
Children's Hospital of Málaga
Avda de los Angeles s/n
E–29011 Málaga (Spain)

0301–0163/95/0449–0018
$8.00/0

Table 1. Variables in the group as a whole

	Average	SD	Range
Compliance	70.0	15.9	65–100
Knowledge	52.9	14.4	18–93
Duration of therapy, years	3.02	1.8	0.4–8.7
Age, years	13.3	2.3	7.9–18.5
Increase SDS stature (n = 88)	0.77	0.74	–0.38–2.87

Table 2. Variables broken down by diagnosis (means ± SD)

	Turner	GH deficiency	Short stature
Compliance	70 ± 15.4	70.5 ± 18.4	69.8 ± 14.4
Knowledge	55 ± 17.3	49.7 ± 18.1	54.6 ± 16.8
Duration of therapy, years	3.1 ± 1.5	2.7 ± 2.1	3.1 ± 1.7
Age, years	12.5 ± 2.2	12.7 ± 2.2	13.6 ± 2.4
Variation SDS stature	1.04 ± 0.7	0.84 ± 0.9	0.66 ± 0.6
n	12	27	49

Table 3. Variables corresponding to little knowledge and poor compliance (means ± SD)

	Knowledge <37.5% (n = 19), %	Compliance <57% (n = 24), %
Compliance	72.1 ± 16.7	
Knowledge		52.2 ± 16.2
Duration of therapy, years	2.8 ± 2.2	2.8 ± 2.0
Age, years	12.0 ± 2.4	13.1 ± 2.5
Variation SDS stature	0.61 ± 0.7	0.69 ± 0.65

Table 4. Variables broken down according to whether the subject attended the weekend workshop (means ± SD)

	Workshop	No workshop
Compliance	75.4 ± 18.4	67.1 ± 16.7
Knowledge	63.7 ± 14.7	48.2 ± 16.2
Duration of therapy, years	3.2 ± 1.7	2.8 ± 1.8
Age, years	13.4 ± 1.8	13.2 ± 2.5
Variation SDS stature	0.69 ± 0.61	0.76 ± 0.76
n	28	61

Of the 137 patients on treatment, 90 filled in the form voluntarily. Ten patients were unable to do so due to social or communication problems, which were in principle not related to the results of the survey. Two questionnaires were discarded as the test had been misunderstood. Of the 88 valid questionnaires, 28 had been completed by children who had attended a weekend workshop on growth problems and specifically on GH treatment organized by our team 6 months earlier. No other unusual educational or informative therapy had been performed in the patients.

The patients had been diagnosed as follows: Turner's syndrome (n = 12), isolated or multiple GH deficiency (n = 27) and idiopathic short stature (n = 49).

The questionnaire contained three types of questions: six were of a general nature (1-5-8-9-10-21), eight referred to the subject's knowledge of the treatment (3-4-6-11-13-14-15-16) and seven to his/her degree of acceptance and compliance (2-7-12-17-18-19-20). For each of the questions about knowledge and compliance, a score was established as follows: 1 point (answers marked with **), or 0.5 points (answers marked with *), for answers that were more or less correct, or more or less positive. The score obtained was expressed as a percentage of the maximum points possible. Five variables were considered in each child: score for compliance and knowledge, duration of treatment, the child's age and variation from the height standard deviation score (SDS) in terms of his/her stature.

The statistical analysis consisted of a test of the comparison of percentages and the analysis of the variance for normal variables. Correlation analysis was assessed using r of Pearson's coefficient.

Results

Tables 1 and 2 show the overall results and a breakdown by diagnosis. The questionnaire and the percent result of the answers to it can be found in the appendix.

Scores for adaptation were significantly higher than for knowledge ($p = 0.0001$). No correlation was found between the two variables ($p = 0.45$, $r = 0.08$), even in the children with the least knowledge (<37.5%, $r = 0.12$, n = 19) and the lowest adaptation (<57%, $r = 0.12$, n = 24; table 3).

We did observe a correlation between knowledge scores and the children's ages ($r = 0.3$, $p = 0.0048$), and logically the duration of treatment ($r = 0.38$, $p = 0.00026$), and the variation from the SDS in terms of their stature ($r = 0.5$, $p = 0.00001$).

When we compared the scores of the children who had attended the workshop and those who had not (table 4), the former achieved significantly higher marks, both in terms of acceptance ($p = 0.03$) and knowledge ($p = 0.008$).

An acute lack of knowledge about the aspects directly related with the injection was observed. Approximately 40% of the children do not know how many units to inject, 85% do not know how many units are contained in each vial, 68% are unaware of the commercial name of the substance and 80% do not know how much solvent to add to the mixture. Half of the patients have no idea of when they will complete their treatment.

Discussion

The difference between the scores for knowledge and compliance, with the latter obtaining higher marks, can be explained in terms of the fact that the questions on the survey about knowledge were more objective and perhaps difficult to answer, while it may have been easier to choose the right answers to the questions about compliance. This may also explain why no correlation was found between these variables. The child's compliance with the treatment is checked primarily by the parents, and this also constitutes an additional motive to score higher.

Our results were similar to those of the English study, where 40% of the children did not know the amount of GH they take. This lack of knowledge is worrying for those who prepare the mixture themselves (between 70 and 80%). We believe that to date the preparations that use the pen system do not solve the problem, since at times it is even more complex to use than the classic syringe.

Although this is only one of many aspects of an ideal educational program for these children, the effect of their participation in a weekend workshop remained 6 months after the event.

A suitable educational program is obviously required since only 40% of the children were shown by medical personnel how to inject the treatment. The validation and update of this type of questionnaire enable an early detection of errors and appropriate educational therapy not only directed toward improving compliance with the treatment, but also toward increasing the motivation and cooperation of the patient.

Appendix

1 How long ago did you begin treatment?

a Less than 1 year
b Between 1 and 2 years
c Between 2 and 3 years
d Between 3 and 4 years
e More than 4 years

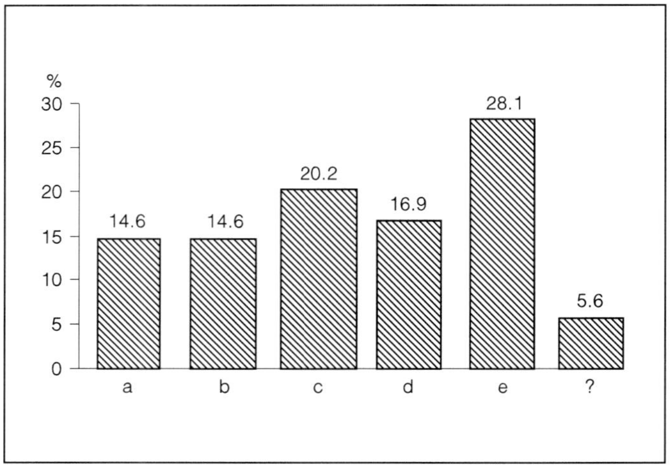

2 Who gives you the shot?

a I do it mayself **
b My father *
c My mother *
d Another member of the family, a friend or neighbor
e A nurse or a doctor

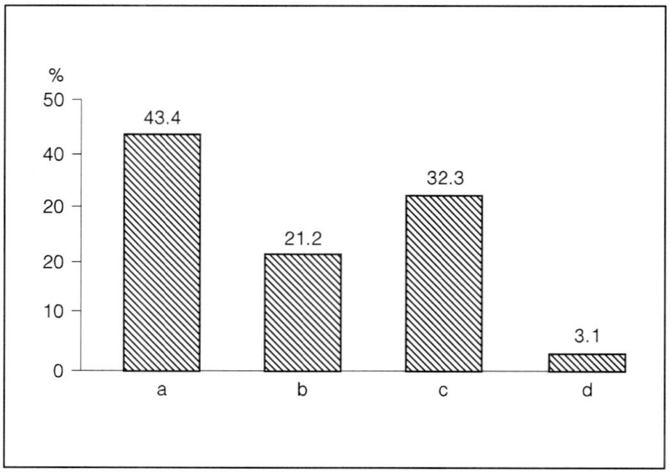

*, ** For explanation, see p. 19.

3 How many units do you take a day?

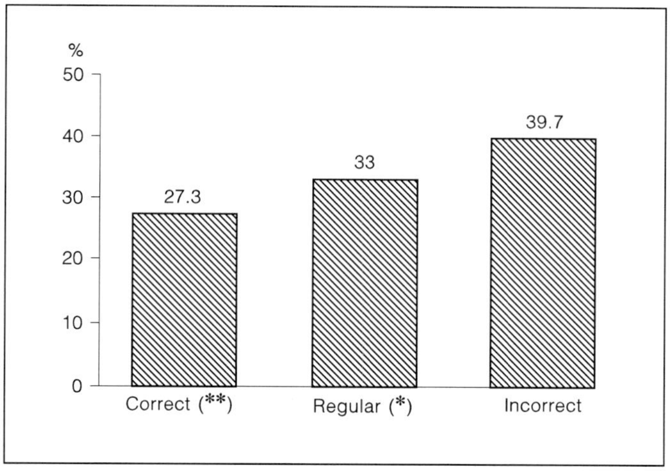

4 How many units of growth hormone are there in each bottle?

a 4
b 6
c 12
d 16
e Another amount
f Trademark

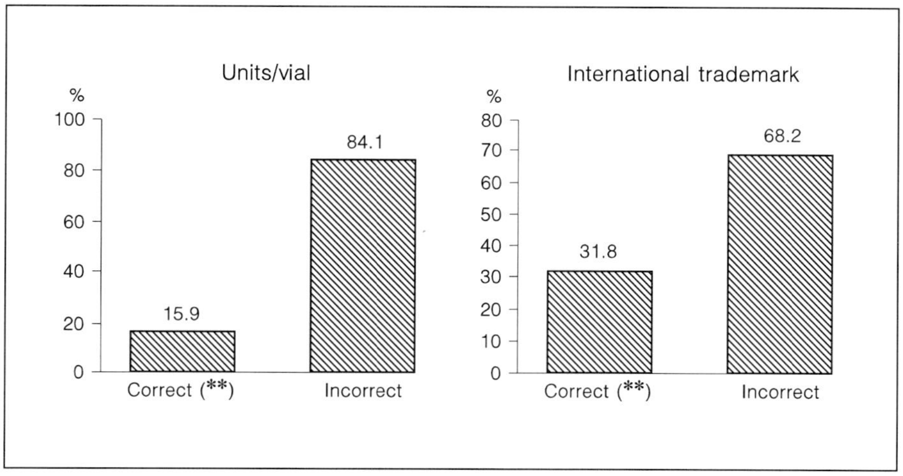

5 Who showed you how to give yourself the shot?

a A doctor at the hospital
b A nurse at the hospital
c My father or my mother
d Someone at the outpatient clinic or health center
e On a weekend workshop
f Nobody

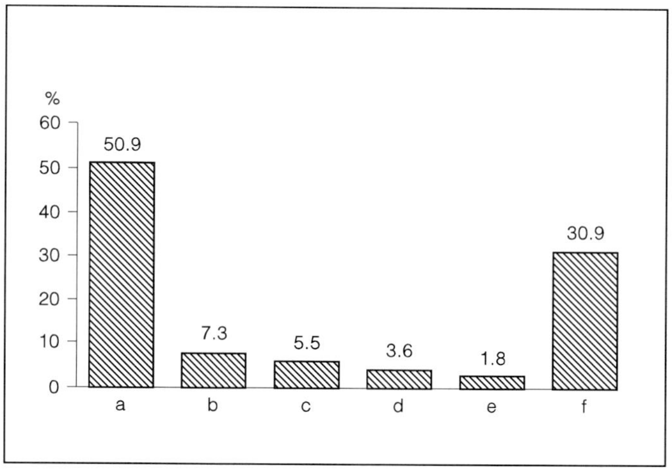

6 How much solvent (water) do you add to the hormone (powder) to prepare the injection?

a 3 ml
b 1 ml
c 2 ml
d It doesn't need mixing
e I don't know
f All the liquid

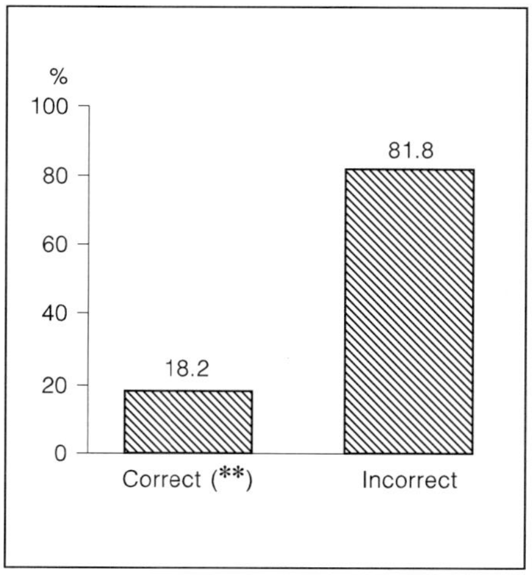

7 Preparing the injection is
- a Very easy **
- b Easy *
- c Complicated
- d Very difficult

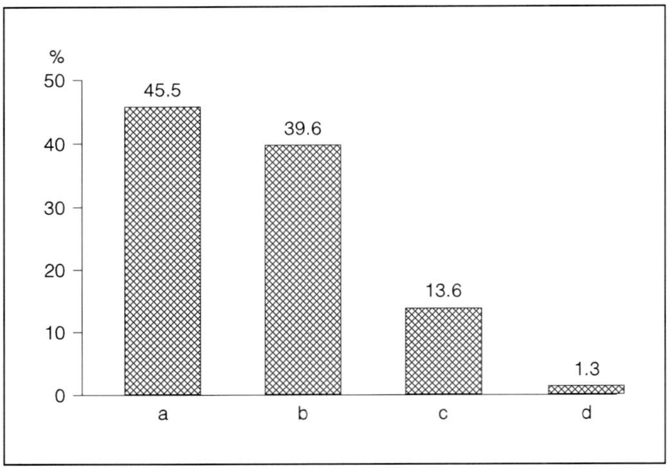

8 Before giving yourself the shot, you clean your skin
- a No
- b With soap and water
- c With alcohol
- d With another antiseptic

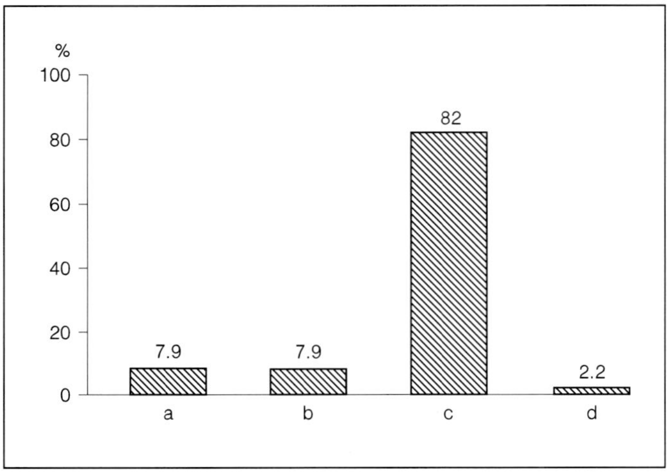

*, ** For explanation, see p. 19.

9 I get the needles and syringes
- a By buying them
- b They come with the medication
- c) They are given to me in the pharmacy
- d They are given to me in the outpatient clinic

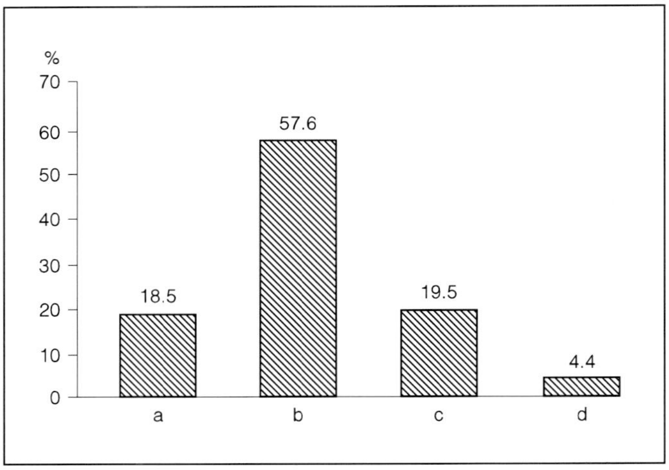

10 The worst or most difficult part of the shot is
- a Preparing the mixture
- b Extracting the amount to be injected
- c Putting the needle under the skin
- d Releasing the liquid under the skin

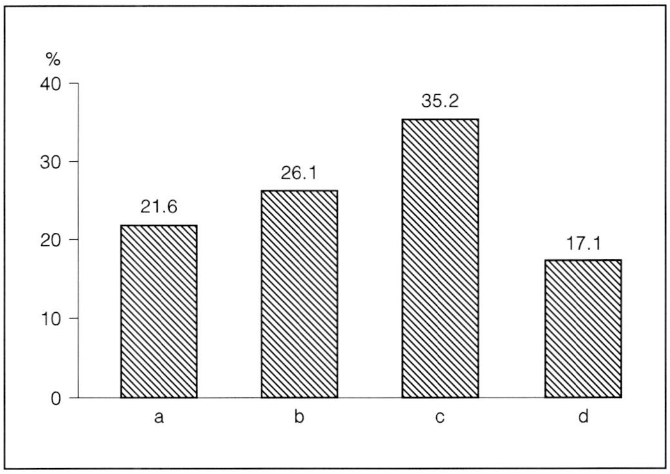

11 Once you have mixed the medication, how many marks of the insulin syringe do you inject every day?

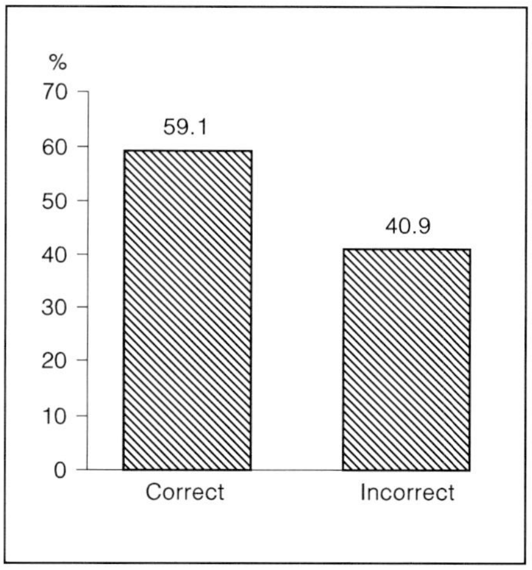

13 Which of these sentences is true?
- a Growth hormone is made in a laboratory
- b Growth hormone is extracted from the human pituitary gland
- c Growth hormone is made with bacteria *
- d Growth hormone is extracted from plants

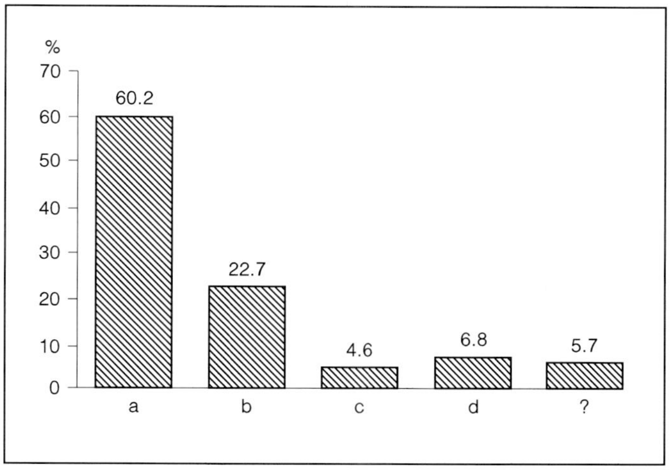

12 Which of these sentences is applicable to you?
- a I give myself a shot almost every day at night
- b I give myself a shot every day when I come back from school
- c I give myself a shot every day at night **
- d I give myself a shot every day except Sunday, after dinner *

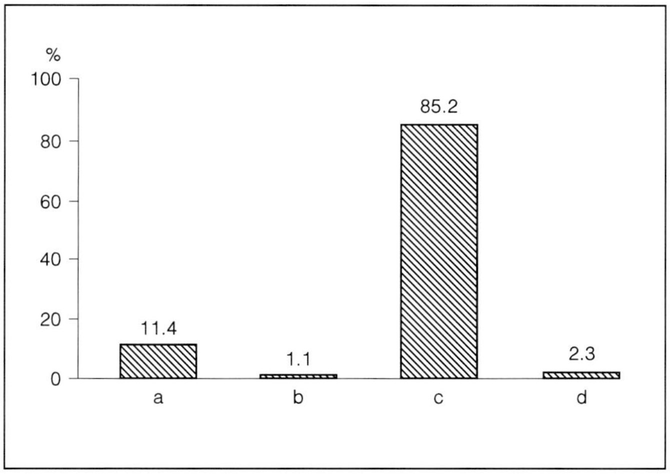

14 Which of these sentences seems true to you?
- a Growth hormone should be used quickly after it has been bought
- b It should always be kept in the lower part of the refrigerator **
- c Once it has been mixed, it lasts 2 weeks
- d When it has been mixed, it looks cloudy

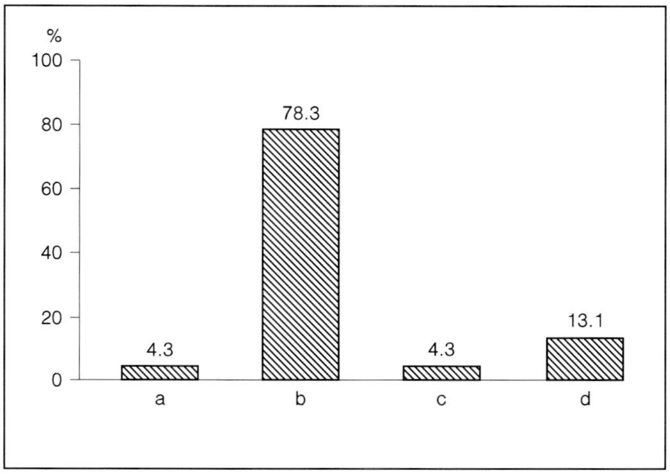

15 I will have to take growth hormone until
- a I am taller than my mother or father
- b I stop growing **
- c I am as tall as the other children of my age
- d I am tired of it

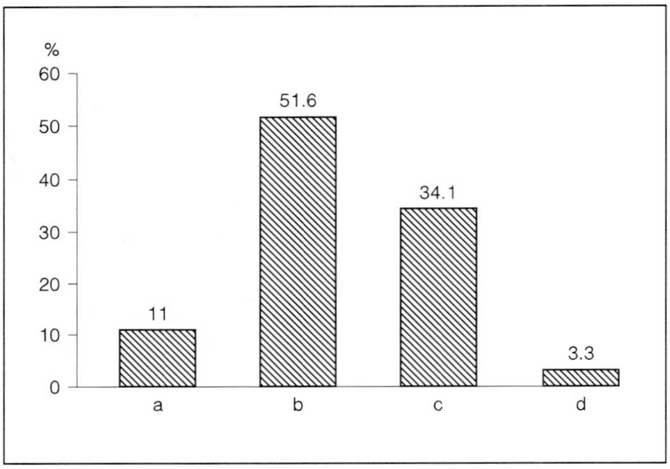

16 I am taking growth hormone because
- a It is helping me to grow
- b I am sick
- c My parents tell me to
- d I get tired

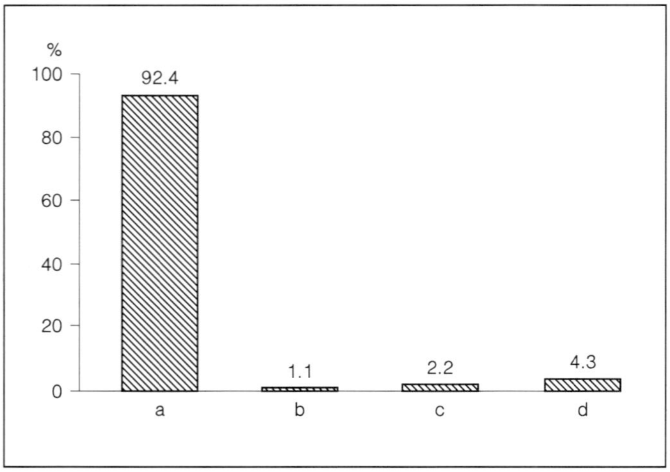

17 When I prepare the injection and there is liquid left over in the ampoule or vial
- a I keep it until the next day **
- b I throw it away
- c I give myself a little more
- d There is never any liquid left over in the bottle

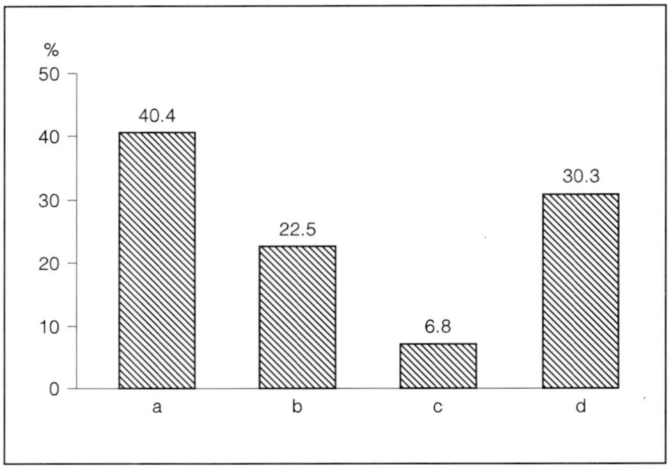

18 Sometimes I forget to take the shot: since my last visit this has happened
- a Less han 5 times
- b From 5 to 10 times
- c More than 10 times
- d Never

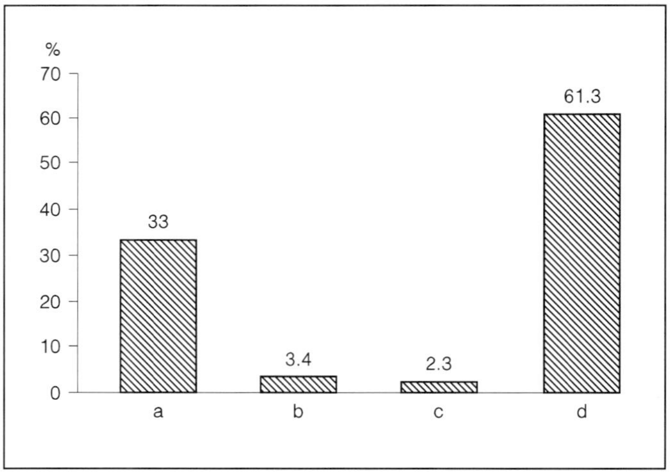

*,** For explanation, see p. 19.

19 Since I have been taking growth hormone
a I have grown a little more *
b I am stronger and eat a lot
c I am much taller **
d I have not noticed anything special

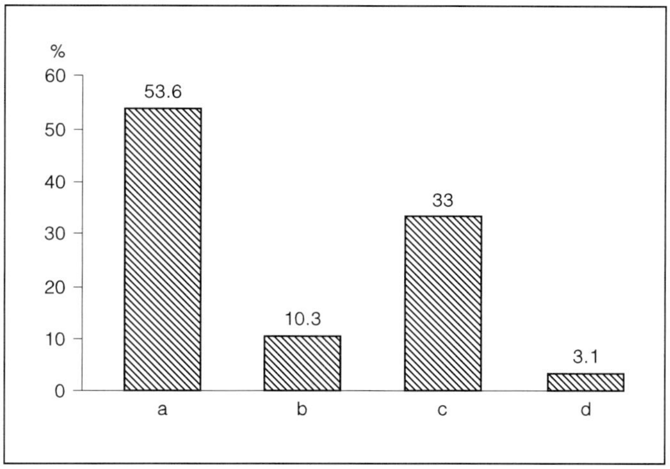

20 If you have to spend a few days away from home
a You stop taking the shots
b You continue taking the shots **
c You use a pen system rather than a syringe *
d You never spend the night away from home

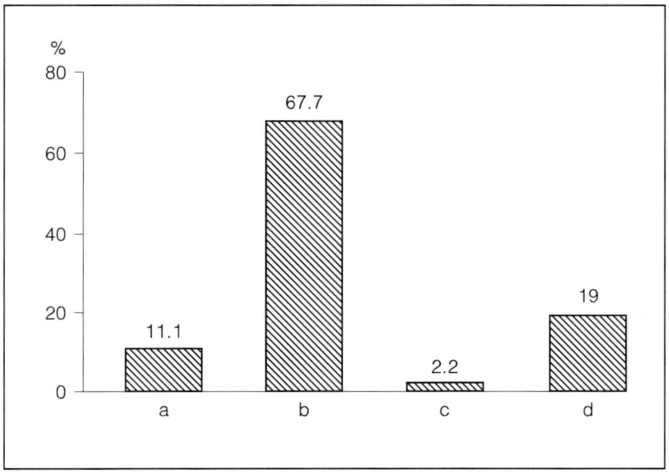

21 Normally I am prescribed growth hormone for
a 1 week
b 2–3 weeks
c 1 month
d More than 1 month

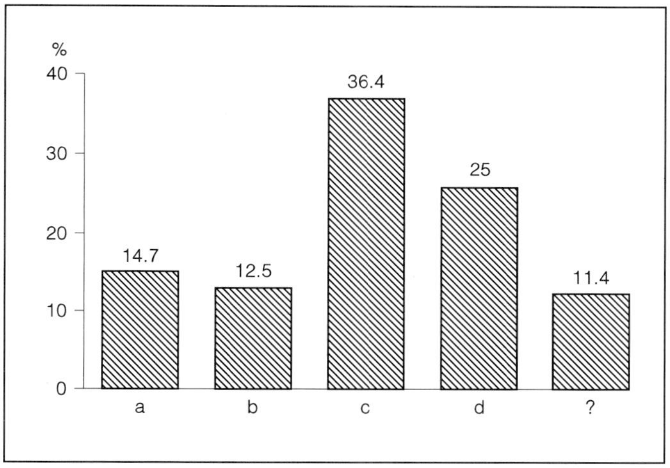

Reference

1 Smith PJ, Hindmarsh PC: Compliance with growth hormone treatment. Are they getting it? Arch Dis Child 1993;68:91–93.

Horm Res 1995;44(suppl 3):26–31

Teresa Arrigo[a]
Filippo De Luca[a]
Sergio Bernasconi[b]
Mauro Bozzola[c]
Luciano Cavallo[d]
Giuseppe Crisafulli[a]
Lucia Ghizzoni[b]
Mohamad Maghnie[c]
Clara Zecchino[d]

Institutes of Pediatrics, Universities of
[a] Messina,
[b] Parma,
[c] Pavia, and
[d] Bari, Italy

Catch-Up Growth and Height Prognosis in Early Treated Children with Congenital Hypopituitarism

Key Words
Growth hormone
Growth hormone deficiency
Congenital hypopituitarism
Catch-up growth
Height prognosis
Early GH treatment

Abstract
The aim of this retrospective study was to ascertain, whether an early growth hormone (GH) treatment can normalize height prognosis of children with congenital GH deficiency (GHD). The study covers 23 children with early onset GHD who received GH 0.1 U/kg/day from the beginning of therapy (0.4–4.9 years). This dose was corrected for weight every 3 months during the whole duration of treatment ($\overline{X}$ 7.9 ± 2.4 years). As a consequence of the significant growth acceleration induced by GH treatment, the patients' height deficiency at the last check had changed from a range of between –8.0 and –1.8 SDS to between +0.5 and –4.3 SDS; on the average, it was significantly less severe than before treatment. Satisfactory growth acceleration was achieved in most patients concomitantly with accelerated bone maturation, as was shown by the stable height age/bone age ratio observed during the follow-up period. The predicted ultimate height was significantly greater than the pretreatment height and it did not differ from the target height. It is concluded that catch-up growth to the target percentile in GHD patients is possible, provided that substitutive treatment is begun during the first years of life and that GH doses are adjusted periodically for weight changes.

Introduction

Full catch-up growth to the target percentile has so far been uncommon in growth hormone (GH)-treated GH-deficient (GHD) children, their catch-up seldom reaching 2 standard deviation scores (SDS) from the start of treatment up to adult height achievement [1, 2]. Moreover, the mean final stature was subnormal, in spite of therapy, in most reported series [3–5]. It has generally been suggested that the disappointing long-term auxological results of GH replacement in GHD patients are due to treatment having been started late [2], though the factors involved have not yet been studied systematically.

The aim of the present retrospective investigation was to ascertain whether an early start of GH treatment, within the first 5 years of life, can normalize the height prognosis of GHD patients.

Prof. Filippo De Luca, MD
Istituto di Clinica Pediatrica
Policlinico Universitario
I-98100 Messina (Italy)

0301–0163/95/0449–0026
$8.00/0

Patients and Methods

The patients studied were those with early onset and nonhereditary GHD who had presented over the previous 15 years, provided that the following criteria were fulfilled: (a) start of GH therapy before the chronological age (CA) of 5 years; (b) treatment duration of more than 3 years; (c) bone age (BA) of at least 6 years at the time of the last check, since Bailey-Pinneau adult prediction [6] cannot be made for 'growth-retarded or average children' with Greulich-Pyle BA [7] of less than 6 years.

The neonatal clinical data of these 23 patients were obtained from their hospital records. In all of them GHD was suspected on the basis of auxological features and was confirmed by a serum GH peak, after two pharmacological stimulation tests (insulin and clonidine), below 10 ng/ml. GH assays were performed by the local hormone laboratories, using commercial kits bases on polyclonal antibodies.

Other thyroid and adrenal evaluations were performed, leading to a classification of isolated GHD (IGHD) or multiple pituitary hormone deficiency (MPHD). Gonadotropin (Gn) secretion was not investigated, as it is not possible to exclude Gn deficiency during the first 5 years of life. In 3 cases, however, Gn deficiency was diagnosed on clinical grounds: the association of microphallus with bilateral cryptorchidism (cases 13 and 21), and the lack of spontaneous puberty (case 23).

All the 23 patients received GH 0.1 IU/kg/day s.c. from the start of therapy (mean age 3.5 ± 1.3 years, range 0.4–4.9). This dose was corrected for weight every 3 months during the whole duration of treatment (7.9 ± 2.4 years, range 3.2–12.3). The patients with MPHD received also hydrocortisone (5 $mg/m^2/day$) and/or *L*-thyroxine (5 μg/kg/day) replacement.

During the follow-up period all the patients underwent 6-monthly evaluations of growth velocity and bone maturation. Height was measured by a wall-mounted Harpenden stadiometer; heights and growth velocities were expressed as SDS from the British standards [8]. BA was graded according to the Greulich-Pyle method [7]. Sexual development was graded using the classification of Tanner [9]. Target height in centimeters was determined by (father's height + mother's + 13)/2 for boys and (father's height + mother's – 13)/2 for girls [10] and expressed as SDS from the adult British standards. Adult height was predicted by the Bailey-Pinneau method [6] at the time of the last check.

The statistical evaluation of the data was performed by means of the Student's t test, both paired and unpaired, Pearson's χ^2 test and correlation analysis; two-tailed p values <0.05 were considered significant. Data are expressed as mean ($\overline{X}$) ± SD.

Results

Clinical and endocrine features of our patients are shown in table 1. Thirteen patients (56.6%) were male. The prevalence of IGHD was 15/23 (65.2%). All the 8 subjects with MPHD exhibited a thyrotropin (TSH) defi-

Table 1. Clinical and endocrine features of 23 patients with early treated GHD

Patient	Sex	Delivery	Neonatal asphyxia	Hypo-glycemia	Micro-penis	Cryptor-chidism	Prolonged jaundice	GH peak	Additional hormone deficiencies
1	F	CS	–	–			–	0.3	TSH
2	F	Normal	–	–			–	0.5	–
3	F	Normal	+	+			–	4.2	–
4	F	Breech	+	+			–	0.5	–
5	F	Normal	–	–			–	8.6	–
6	F	CS	+	–			–	5.6	–
7	F	CS	+	–			–	7.9	–
8	F	Normal	–	–			–	4.4	–
9	F	Normal	–	–			–	1.7	–
10	F	Normal	–	–			–	8.1	–
11	M	Normal	–	–	–	–	–	0.6	TSH
12	M	Breech	–	–	–	–	–	5.3	–
13	M	CS	+	+	+	+	+	2.5	TSH, ACTH, Gn
14	M	Normal	–	–	–	–	–	4.3	–
15	M	Normal	–	–	+	–	–	0.9	TSH
16	M	CS	+	–	–	–	–	3.3	–
17	M	Normal	–	+	+	–	+	1.7	TSH
18	M	Normal	–	–	–	–	–	5.3	–
19	M	Normal	–	–	–	–	–	4.3	–
20	M	Normal	–	–	–	–	–	0.9	–
21	M	Breech	–	+	+	+	–	0.7	TSH, Gn
22	M	Breech	+	–	–	–	–	1.5	TSH
23	M	Breech	–	–	+	–	+	0.9	TSH, ACTH, Gn

ciency (34.8%) associated with corticotropin (ACTH) deficiency in 2 cases (8.7%) and/or with Gn deficiency in 3 cases (13%).

The delivery was reported as normal in 13/23 infants (56.6%), cesarean section (CS) in 5 cases (21.7%) and breech delivery in the remaining 21.7%. Neonatal asphyxia, documented by an Apgar score at 5 min of 0–7, was reported in 7 infants (30.4%). A perinatal insult (breech delivery and/or neonatal asphyxia) was recorded in 10 patients (43.5%). Clinical signs documenting the existence of pituitary dysfunction in utero or immediately after birth (hypoglycemia, micropenis, bilateral cryptorchidism, prolonged jaundice) were observed in 7 infants (30.4%) (table 1). The age at the start of GH replacement was practically the same in these 7 patients with MPHD as in the remaining 16 (3.5 ± 1.0 vs. 3.5 ± 1.4 years). Moreover, the age at the start of therapy was similar in the patients with IGHD (3.7 ± 1.1 years) and in those with MPHD (3.1 ± 1.6 years). Seventeen subjects (73.9%) exhibited a GH peak in response to both pharmacological stimulations <5 ng/ml. In 8 cases (34.8%) this peak was even lower than 1 ng/ml (table 1).

At the time of starting GH treatment, 21/23 patients (91.3%) showed a height retardation which exceeded 2 SDS (table 2). In 2 patients (cases 13 and 21) with less marked height retardation (–1.8 SDS) early diagnosis of hypopituitarism was suggested by the coexistence of other

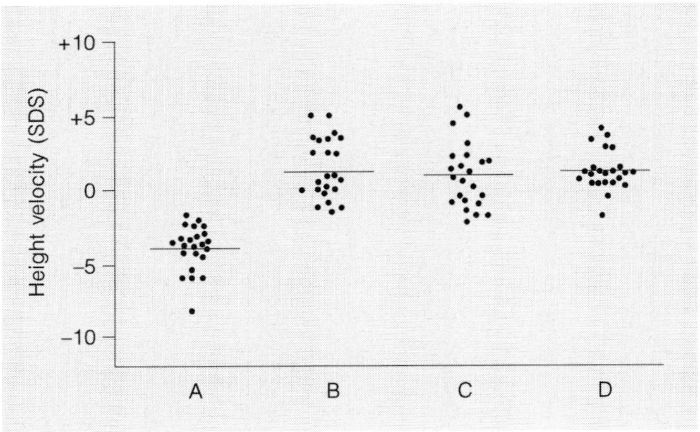

Fig. 1. Individual and average (—) values of height velocity in the whole series at the start of GH treatment (A) and after the first (B), second (C) and third year of therapy (D).

Table 2. Patients' auxological data at the time of GH treatment institution

Patient	CA years	BA years	BA/CA ratio	HA/BA ratio	Height (SDS)	Target height (SDS)
1	0.4	0.1	1.0	1.0	–7.0	–0.8
2	1.7	0.5	0.3	1.0	–5.6	–0.4
3	2.4	0.7	0.3	1.4	–5.0	+1.5
4	3.8	1.2	0.3	0.6	–8.0	–2.3
5	4.4	2.6	0.6	1.6	–2.5	–1.3
6	4.6	2.0	0.4	1.0	–2.2	–0.5
7	4.6	2.6	0.6	1.1	–2.1	–2.0
8	4.8	2.3	0.5	0.9	–4.8	–1.3
9	4.9	3.0	0.6	0.8	–5.0	–0.2
10	4.9	2.0	0.4	1.3	–3.3	–0.8
11	1.6	0.3	0.2	1.0	–7.1	–0.7
12	2.1	0.5	0.2	2.6	–2.3	–0.2
13	2.2	2.0	0.9	0.9	–1.8	–1.0
14	2.7	1.3	0.5	1.3	–2.8	–0.8
15	3.3	0.5	0.2	2.6	–5.1	–2.2
16	3.4	2.0	0.6	0.8	–4.0	–2.0
17	3.5	0.8	0.2	1.3	–5.8	–0.5
18	3.7	2.0	0.5	1.1	–2.8	–1.0
19	3.9	1.7	0.4	1.7	–2.1	–0.9
20	4.0	1.5	0.4	1.5	–2.8	–0.6
21	4.5	2.0	0.4	1.7	–1.8	0
22	4.5	2.5	0.6	1.0	–3.1	–1.6
23	4.9	1.5	0.3	1.4	–4.5	+0.8
$\overline{X}$ ± SD	3.5±1.3	1.5±0.8	0.45±0.21	1.29±0.51	–4.0±1.9	–0.8±0.9

hypopituitary features, such as hypoglycemia, microphallus and cryptorchidism.

At the beginning of GH therapy, BA was severely delayed in most of our patients, as indicated by the very low BA/CA ratio (0.45 ± 0.21, range 0.2–1.0) (table 2). The average height age (HA)/BA ratio was 1.29 ± 0.51 (range 0.6–2.6) (table 2). Target height was within normal range (i.e. ± 2 SDS) in all but 2 patients (cases 4 and 15) (table 2). Pretreatment height velocity was severely impaired in almost all the patients ($\bar{X}$ −4.0 ± 1.5 SDS, from −1.8 to −8.2) (fig. 1). The start of GH therapy was followed during the first year by a significant growth acceleration ($\bar{X}$ 1.4 ± 2.1 SDS, $2p < 0.0001$), which persisted during both the second (0.7 ± 2.3 SDS, $2p < 0.0001$) and the third year of treatment (1.0 ± 1.4 SDS, $2p < 0.0001$) (fig. 1).

Patients' height at the last examination was between 0.5 and −4.3 SDS, as compared with between −8.0 and −1.8 SDS at diagnosis. The average height at the last check was −1.5 ± 1.3, as against −4.0 ± 1.9 SDS ($2p < 0.0001$) before treatment. The average statural recovery during the follow-up period in all patients was 2.5 ± 2.1 SDS (0.33 ± 0.24 SDS/year of treatment).

The patients' height at the last check exceeded the target height in 10/23 subjects (cases 1, 4, 6, 7, 9, 13, 15, 16, 19 and 22) and on the average was very similar to it (−1.5 ± 1.3 vs. −0.8 ± 0.9 SDS). Height deficiency exceeded 2 SDS in only 6/23 cases (vs. 21/23 at the beginning of treatment; $\chi^2 = 20.1$, $p < 0.0005$) (table 3). Three of these 6 patients with subnormal height at the last examination (cases 8, 11 and 17) had exhibited a satisfactory catch-up growth (between 1.8 and 3.8 SDS), whilst in the remaining 3 (cases 10, 20 and 23) no response to treatment was seen.

No significant relationship was found among height before treatment, height at the last check and target height.

The satisfactory catch-up growth observed in most patients was achieved concomitantly with accelerated bone maturation from the first to the last check (BA/CA ratio 0.86 ± 0.14 vs. 0.45 ± 0.21, $2p < 0.0001$), as is indicated by the stable HA/BA ratio (1.02 ± 0.14 vs. 1.29 ± 0.51) (table 3).

Table 3. Patients' auxological data at the time of the last evaluation

Patient	CA years	BA years	BA/CA ratio	HA/BA ratio	Height (SDS)	Predicted height (SDS)	Puberty stage	Puberty onset years
1	8.4	7.5	0.9	1.1	−0.2	0	P1 B1	–
2	8.3	6.0	0.7	1.2	−1.5	+0.5	P1 B1	–
3	11.2	6.5	0.6	1.4	−2.0	+2.4	P1 B1	–
4	11.6	11.5	1.0	1.0	−0.3	−0.8	P3 B4	9.5
5	7.6	6.0	0.8	1.0	−1.6	−1.0	P1 B1	–
6	11.0	10.0	0.9	1.1	−0.3	0	P1 B1	–
7	12.1	9.7	0.8	1.1	−1.0	+0.9	P1 B1	–
8	15.2	15.0	1.0	0.8	−3.0	−3.0	P3 B4	12.7
9	14.5	16.0	0.9	0.9	+0.2	+0.1	P5 B5	11.5
10	11.0	7.4	0.7	1.1	−3.0	0	P1 B1	–
11	9.1	6.2	0.7	0.9	−3.3	−1.3	P1 G1	–
12	10.9	8.0	0.7	1.1	−1.8	+0.2	P1 G1	–
13	10.8	11.5	1.1	1.1	+0.4	+1.0	P1 G1	–
14	12.0	12.8	1.1	0.8	−1.7	−2.8	P1 G1	–
15	10.6	7.5	0.7	1.1	−2.0	+0.2	P1 G1	–
16	13.6	12.5	0.9	0.9	−1.5	−0.8	P3 G3	11.0
17	15.8	17.8	1.1	0.8	−3.1	−3.2	P5 G5	11.6
18	8.2	6.2	0.8	1.1	−1.7	−0.4	P1 G1	–
19	9.3	9.0	1.0	1.0	0	+0.6	P1 G1	–
20	14.5	12.0	0.8	1.0	−3.0	+0.1	P2 G3	11.1
21	9.0	8.5	0.9	1.0	−0.7	0	P1 G1	–
22	11.2	10.5	0.9	1.1	+0.5	+1.4	P1 G1	–
23	17.2	13.8	0.8	0.9	−4.3	−2.1	P3 G3	16.5
$\bar{X}$ ± SD	11.4±2.6	10.0±3.3	0.86±0.14	1.02±0.14	−1.5±1.3	−0.3±1.4		

At the last check, 2 patients (cases 9 and 17) had reached their adult height and 5 others were not very far from it, having entered spontaneous (cases 4, 8, 16 and 20) or testosterone-induced puberty (case 23). Predicted ultimate height significantly improved with respect to the pretreatment height ($2p < 0.0001$) and was slightly even though not significantly higher than target height (–0.3 ± 1.4 vs. –0.8 ± 0.9 SDS). It did not correlate with either CA or BA at the start of therapy or with height before treatment or target height. In only 4 patients (cases 8, 14, 17 and 23) was the predicted adult height subnormal, i.e. ≤2 SDS (from –2.1 to –3.2) (table 3).

Discussion

Although GH treatment in GHD children has been employed for more than three decades, the reasons for the disappointing results in terms of adult height [1, 3–5] have not been identified yet.

In the recent literature it is frequently debated whether or not a pharmacological delay in the onset of puberty or a prolongation of puberty can improve the statural outcome of GHD children with poor height prognosis [11]. It is also discussed whether or not GH doses should be increased at the entrance into puberty in order to optimize the auxological performance, without accelerating puberty progression [12–14]. The optimal modalities and doses of sex hormone replacement in GHD children without spontaneous onset of puberty are also discussed [12, 15]. Most investigators, however, agree that 'the battle for a normal height needs to be won before puberty' [14] and that during puberty – spontaneous or induced – height prognosis cannot substantially improve. The final height of GHD patients is mainly conditioned by two factors: (a) height prognosis at the start of GH substitution [3, 4]; (b) height prognosis at the start of puberty [5, 12, 15]. Consequently, full catch-up growth in hypopituitary patients first of all needs early diagnosis and early treatment [2, 16]. Another important goal is to improve height prognosis at the start of puberty, which can be achieved by optimizing treatment modalities (daily GH injections, periodical adjustment of doses in relation to weight increase).

The effectiveness of a management strategy based on early diagnosis and optimal prepubertal treatment is confirmed by the results of the present study. Thanks to early diagnosis, our patients' height prognosis at the beginning of GH replacement was not severely impaired, as is shown by the positive HA/BA ratio found in most of them. Moreover, the optimization of GH therapy with daily subcutaneous injections and 3-monthly adjustments of dose for weight enabled most patients to achieve full catch-up growth to the target percentile. In fact, at the time of the last check, the average height was very close to the target height and even higher than that in 44% of patients. Likewise, predicted height did not differ on the average from target height and exceeded it in 78% of patients.

Although the importance of treating GHD infants early was first pointed out 7–9 years ago [16–18], the recent literature includes few papers on this topic [19–22]. Even without long-term studies, the authors of these concluded, as we do, that GH treatment in GHD patients is more effective in normalizing growth velocity and height when it is begun early. Moreover, the best response to GH therapy seems to be in young children with severe idiopathic GHD who receive daily weight-adjusted doses [22].

This means that most GHD children should receive daily injections for 10–12 years, which is a heavy load for the child and his family. Nevertheless, if the aim of GH therapy in GHD is to achieve a final height in the most satisfactory range, i.e. very near to or above the target height, the diagnosis must be made as early as possible. Some clinical data may be helpful for early detection of hypopituitarism, as in the present series: history of perinatal insults, neonatal hypoglycemia and/or prolonged jaundice; micropenis and/or cryptorchidism. It is to be underlined that GHD in early onset hypopituitarism is frequently very severe, as is demonstrated by the blunted GH responses to pharmacological stimulations in many of our patients, and those reported elsewhere [21]. The severity of GHD does not seem to have a negative impact on growth response to GH therapy, as indicated by the inverse correlation recently found between the maximum stimulated serum GH and height velocity in the first year of treatment [22].

In conclusion, the results of our retrospective study demonstrate that complete catch-up growth to the target percentile in GHD patients is possible, provided that substitutive treatment is begun during the first years of life and that GH doses are periodically adjusted for weight changes.

References

1 Job JC, Joab N, Toublanc JE, Canlorbe P: Résultats à terme des traitements par l'hormone de croissance humaine. Arch Fr Pédiatr 1984;41:477–482.

2 Job JC: Early diagnosis and early treatment of growth hormone deficiency. Horm Res 1989; 31:149–152.

3 Burns EC, Tanner JM, Preece MR, Cameron N: Final height and development in 55 children with idiopathic growth hormone deficiency, treated for between 2 and 15 years with human growth hormone. Eur J Pediatr 1981;137:155–164.

4 Joss E, Zuppinger K, Schwarz HP, Roten H: Final height of patients with pituitary growth failure and changes in growth variables after long-term hormonal therapy. Pediatr Res 1983; 17:676–679.

5 Bourguignon JP, Vandeweghe M, Vanderschueren-Lodeweyckx M, Malvaux P, Wolter R, Du Caju M, Ernould C: Pubertal growth and final height in hypopituitary boys: A minor role of bone age at onset of puberty. J Clin Endocrinol Metab 1986;63:376–382.

6 Bailey N, Pinneau SR: Tables for predicting adult height from skeletal age: Revised for use with the Greulich-Pyle hand standard. J Pediatr 1952;40:423–441.

7 Greulich WW, Pyle SI: Radiographic atlas of skeletal development of the hand and wrist. Stanford, Stanford University Press, 1959.

8 Tanner JM, Whitehouse RH: Clinical longitudinal standards for height, weight, height velocity, weight velocity and the stages of puberty. Arch Dis Child 1976;51:170–179.

9 Tanner JM: Growth at Adolescence. Oxford, Blackwell Scientific, 1962.

10 Tanner JM, Goldstein H, Whitehouse RH: Standards for children's height at age 2 to 9 years allowing for height of parents. Arch Dis Child 1970;45:755–762.

11 Hibi I, Tanaka T, Tanae A, Kagawa S, Hashimoto N, Yoshizawa A, Shizume K: The influence of gonadal function and the effect of gonadal suppression treatment on final height in growth hormone (GH)-treated GH-deficient children. J Clin Endocrinol Metab 1989;69: 221–226.

12 Ranke MB, Butenandt O: Idiopathic growth hormone deficiency: Final height to treatment with growth hormone and effects of puberty and sex steroids; in Frisch H, Laron Z (eds): Induction of Puberty in Hypopituitarism. Rome, Ares-Serono Symposia, 1988, pp 85–99.

13 Darendelier F, Himdmarsh PC, Preece MA, Cox I, Brook CGD: Growth hormone increases rate of pubertal maturation. Acta Endocrinol 1990;122:414–416.

14 Ranke MB, Lindberg A: Pubertal growth in patients with idiopathic growth hormone deficiency. Kabi Int Growth Study 1993;10:27–36.

15 Bourguignon JP: Linear growth as a function of age at onset of puberty and sex steroid dosage: Therapeutic implications. Endocrinol Rev 1988;9:467–488.

16 Josefsberg Z, Bauman B, Pertzelan A, Laron Z: Greater efficiency of human growth hormone therapy in children below five years of age with growth hormone deficiency. A 5-year follow-up study. Horm Res 1987;27:126–133.

17 Mugnier E, Ployard F, Roy MP, Fermanin J, Rappaport R: Résultats du traitement par l'hormone de croissance humaine chez les enfants insuffisants hypophysaires âgés de moins de 7 ans. Arch Fr Pédiatr 1985;42:671–676.

18 Job JC, Chaussain JL, Garnier P, Rolland A, Joab N: Dose-response relationship in the treatment of hypopituitary children with human growth hormone: A retrospective survey. Acta Paediatr Scand 1987;337(suppl):93–105.

19 Arrigo T, Ghizzoni L, Volta C, Lombardo F, Bernasconi S, De Luca F: Crescita di recupero e prognosi staturale in pazienti con deficit di GH trattati fin dai primi 5 anni di vita. Riv Ital Pediatr 1992;18:326–329.

20 Di Maio S, Salerno MC, Argenziano A, Aloi G, Gasparini N, Sandomenico ML, Franzese A, Tenore A: Trattamento precoce dell'ipotuitarismo. Riv Ital Pediatr 1994;20:21–26.

21 Mugnier E, Rappaport R: Déficit précoce en hormone de croissance. Résultat du traitement par hGH et causes d'échec. Seminaire d'Endocrinologie Pédiatrique et Diabétologie de l'Hôpital des Enfants Malades, Paris, 1994, pp 7–13.

22 Blethen SL, Compton P, Lippe BM, Rosenfeld RG, August GP, Johanson A: Factors predicting the response to growth hormone (GH) therapy in prepubertal children with GH deficiency. J Clin Endocrinol Metab 1993;574:574–579.

Horm Res 1995;44(suppl)3:32–34

Atilla Büyükgebiz

Department of Pediatric Endocrinology,
Dokuz Eylül Medical Faculty,
İzmir, Turkey

Treatment of Constitutional Delayed Puberty with a Combination of Testosterone Esters

Key Words
Constitutional delayed puberty
Testosterone treatment
Predicted adult height

Abstract
Thirteen boys who had constitutional delayed puberty (CDP) were treated with a combination of short- and long-acting testosterone esters (testosterone propionate, testosterone phenylpropionate, testosterone isocaproate). Mean age at the onset of treatment was 14.9 ± 0.6 years and bone age delay was –2.7 ± 0.9 years. An intramuscular dose of 200 mg testosterone was administered 4 times at 3-week intervals and the treated CDP boys were followed for 2 years. All boys with CDP entered puberty after the last dose (testicular volume ⩾4 ml) and growth rate increased from 4.5 ± 0.5 cm/year pretreatment to 8.4 ± 1.6 cm/year posttreatment after the 2-year follow-up period. Height for bone age SD score did not alter significantly from a mean of –1.1 pretreatment to –1.3 posttreatment as well as predicted height pretreatment (173.5 ± 6.6 cm) and posttreatment (173.3 ± 4.9 cm). A combination of testosterone esters in a given dose and schedule is a safe and effective treatment for prepubertal boys with CDP.

Introduction

Constitutional delayed puberty (CDP) is the most common cause of short stature and delayed development in adolescent boys [1]. These boys have delayed sexual maturation, delayed skeletal bone ages and are shorter than age-matched controls [1–3]. For psychological reasons, different drugs have been used for pubertal advancement and short-term growth increment in these adolescents with CDP [2–5]. One of these drugs, used widely for CDP, is testosterone enanthate, a long-acting testosterone ester, and is found to be effective and safe especially in small doses [5–7].

A combination of short- and long-acting testosterone esters (20 mg testosterone propionate, 40 mg testosterone phenylpropionate, 40 mg testosterone isocaproate in a 100-mg ampule) was used in our CDP patients, who were longitudinally followed for 2 years in an attempt to investigate the physiological responses.

Patients and Methods

Thirteen boys aged between 14.2 and 15.4 years were referred to our pediatric endocrinology clinic for evaluation of short stature or delayed puberty. They fulfilled the criteria for CDP (no endocrine or systemic disease, height below third percentile, testicular volume below 4 ml) and all had symptoms of psychological disturbance. 200 mg of testosterone esters were administered intramuscularly 4 times at 3-week intervals and the patients were then followed up for a 2-year period. Clinical data from 13 patients with CDP are shown in table 1. Statistical analysis was performed by t test for matched pairs and Wilcoxon signed ranks test where appropriate. Values of $p < 0.05$ were accepted as statistically significant.

Atilla Büyükgebiz, MD
Department of Pediatric Endocrinology
Dokuz Eylül Medical Faculty
İzmir (Turkey)

0301–0163/95/0449–0032
$8.00/0

1 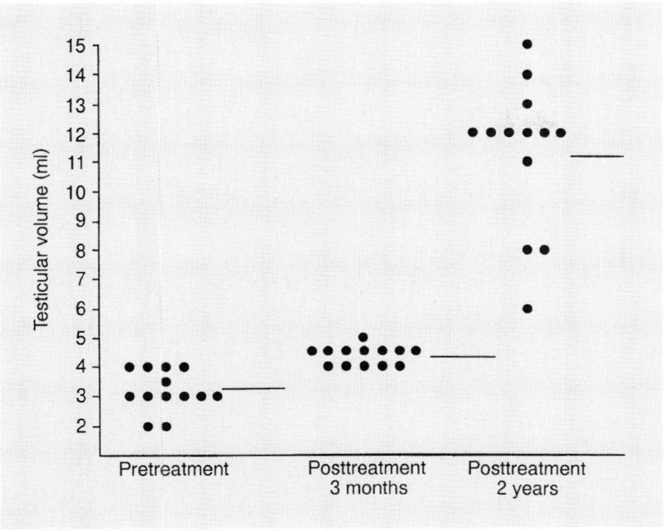

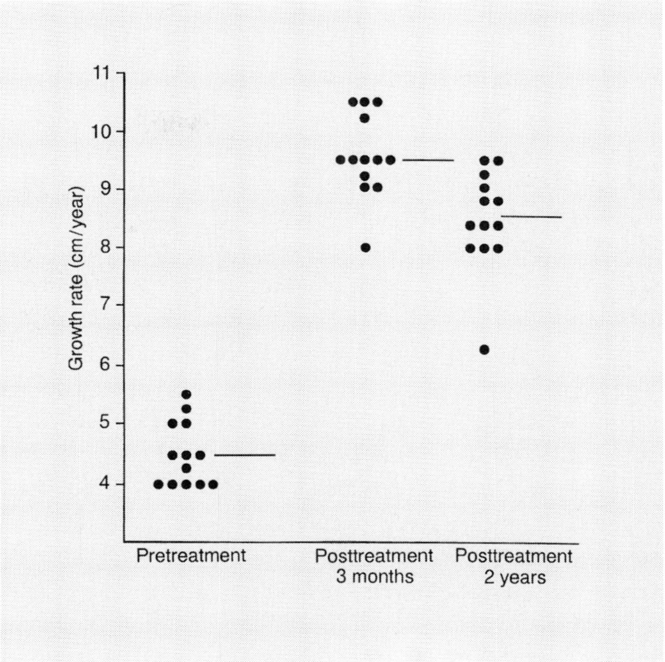 2

Fig. 1. Effect of testosterone therapy on testicular volume in CDP boys. Horizontal line represents the mean value.

Fig. 2. Effect of testosterone therapy on growth rate in CDP boys. Horizontal line represents the mean value.

Table 1. Data for subjects treated with testosterone

	Pretreatment	Posttreatment	
		3 months	2 years
Chronological age, years	14.9±0.6		16.9±0.6
Bone age delay, years	−2.7 (−3.8 to 0.1)		
Height, cm	147.6±7.7	150.9±8.8	164.4±7.7
Growth rate, cm/year	4.5±0.5	p<0.001	8.4±1.6
Predicted adult height, cm	173.5±6.6	p>0.5	173.3±4.9
Height SDS for BA	−1.1 (−2.7 to +1.5)	p>0.5	−1.3 (−3.2 to +1.4)
Testicular volume, ml	3.2±0.7	4.3±0.4	11.3±2.5
Pubic hair (Tanner stage)	1	2	4
CA/BA	1.24		1.18

Growth was measured by using a Harpenden stadiometer and bone age was measured by the method of Tanner et al. [8]. Testicular volume was assessed by using a Prader orchidometer by the same observer and predicted adult height was assessed by the TW2 method [8, 9].

Results

After 3 months of testosterone therapy all 13 CDP boys entered puberty (testicular volume 4.3 ± 0.4 ml and pubic hair stage 2) and no regression was detected after the 2-year follow-up period (fig. 1). Growth rate, which was 4.5 ± 0.5 cm/year, had increased to 8.4 ± 1.6 cm/year after 2 years of follow-up ($p < 0.001$) (fig. 2).

Height SD score for bone age was not significantly different between the pre- and posttreatment period of 2 years ($p > 0.5$). Predicted adult height, which is very important to consider in testosterone treatment, did not vary significantly pretreatment (173.5 ± 6.6) and posttreatment (173.3 ± 4.9) of testosterone ($p > 0.5$). None of the boys experienced adverse reactions during the course of treatment.

Discussion

The child with CDP will spontaneously enter puberty between 14 and 18 years of age and will ultimately achieve a normal adult height. In this normal growth variant situation an approach to the patient and his parents with reassurance and a policy of watchful waiting is necessary. But there are some situations in which the child is so upset by the combination of short stature and delayed appearance of secondary sexual characteristics, that the possibility of intervention needs to be seriously considered [7].

The retardation of growth and delay in sexual maturation exhibited by adolescents have psychological and social consequences. These boys are under the pressure of being short and sexually immature. Some authors have reported that late-maturing boys show greater evidence of negative self-concepts, prolonged dependency needs, and feelings of rejection and rebellion [7, 10]. The same authors claimed that after treatment of CDP, the same group of adolescents demonstrated an improvement in both school and extracurricular social activity, suggesting that normalization of growth and sexual maturation may well permit more effective involvement in normal adolescent behavior [7, 10].

Although it has been shown in several studies that a short course of long-acting testosterone esters promotes puberty in CDP patients without any decrease on predicted adult heights, this is dose-dependent and still a point of investigation.

In our CDP boys treated with a short course of a combination of short- and long-acting testosterone esters, all had entered puberty after the last dose of treatment. There was no significant decrease in height for bone age SD scores and predicted heights after 2 years of follow-up, but growth rate increased significantly. We propose that a combination of testosterone esters (Sustanon) at a dose of 200 mg 4 times at 3-week intervals is as effective as testosterone enanthate at the same dose and schedule used by several authors for CDP [3, 5, 7]. Although there was no significant difference in predicted adult heights, final adult height measurements should be determined in order to make a precise decision.

References

1 Styne DM: Delayed puberty; in Hintz RL, Rosenfeld RG (eds): Growth Abnormalities. New York, Churchill Livingstone, 1987, pp 141–165.
2 Kaplowitz PB: Diagnostic value of testosterone therapy in boys with delayed puberty. Am J Dis Child 1989;143:116–120.
3 Wilson DM, Kei J, Hintz RL, Rosenfeld RG: Effects of testosterone therapy for pubertal delay. Am J Dis Child 1988;142:96–99.
4 Donaldson MDC, Savage DCL: Testosterone therapy in boys with delayed puberty (abstract). Pediatr Res 1988;23:125A.
5 Martin MM, Martin ALA, Mossman KL: Testosterone treatment of constitutional delay in growth and development: Effect of dose on predicted versus definite height. Acta Endocrinol 1986(suppl 279):147–152.
6 Blethen SL, Gaines S, Weldon V: Comparison of predicted and adult heights in short boys: Effect of androgen therapy. Pediatr Res 1984; 18:467–469.
7 Rosenfeld RG, Northcraft GB, Hintz RL: A prospective, randomized study of testosterone treatment of constitutional delay of growth and development in male adolescents. Pediatrics 1982;69:681–687.
8 Tanner JM, Whitehouse RH, Cameron CN, Marshall WA, Healy MJR, Goldstein H: Assessment of skeletal maturity and prediction of adult height (TW2 method), ed 2. New York, Academic Press, 1983.
9 Zachmann M, Prader A, Kind HP, Haflinger H, Budliger H: Testicular volume during adolescence. Helv Pediatr Acta 1974;29:61–72.
10 Mussen PH, Jones MC: Self-conceptions, motivations and interpersonal attitudes of late and early-maturing boys. Child Dev 1957;28:243.

Horm Res 1995;44(suppl 3):35–41

Michael B. Ranke

University Children's Hospital, Tübingen, FRG

Growth Hormone Therapy in Turner Syndrome

Analysis of Long-Term Results

Key Words
Growth hormone therapy
Turner syndrome
Height, final

Abstract
The favorable long-term results of growth hormone (GH), alone or in combination with oxandrolone, reported by an American study [J Pediatr 1992;121:49–55], have so far not been confirmed by other investigators. However, from the presently available data it can be deduced that with higher than substitutive doses of GH, height can be improved during childhood and will be normalized in many patients with Turner syndrome. The future task must be to optimize treatment with respect to the dose and timing of GH (and oxandrolone) in individuals in order to ensure optimal statural outcome at the lowest risk and material costs.

Introduction

Turner syndrome (TS) is a rather frequent disorder (approx. 1:2,500 liveborn females) which is characterized by three main clinical features: (1) abnormal external appearance and abnormalities of some internal organs; (2) malformation of the ovaries, and (3) short stature during childhood and in adult life. Deviation of the appearance and abnormalities of internal organs varies. In some cases the characteristic features allow the diagnosis 'prima vista' and as early as in the neonatal phase. In other cases the symptoms are more subtle and need to be sought for. Gonadal dysgenesis also varies to a great extent. It tends to progress with age, thus allowing a fraction of patients to enter puberty, and even allows fertility in rare cases. Since short stature is the most constant finding in TS, its treatment appears to be of great importance. In recent years, attempts have been made to improve height in TS with growth hormone (GH). The goal of this review is to assess the long-term results of such therapy. Since, however, the evaluation of the response depends so much on the specifics of spontaneous growth in this disorder, some methodological aspects are also discussed.

Spontaneous Growth in TS

In 1938, Henry H. Turner [1] described a disorder in 7 females characterized by sexual infantilism, congenital webbed neck, cubitus valgus and short stature. As early as 1930, Otto Ullrich [2] had recognized the same disorder as a specific entity. In the light of recent interest in the spontaneous growth of patients with TS, it is of interest to note that Otto Ullrich's patient reached an adult height of 144.6 cm [3]. Today it is well established that short stature is the most constant of all features in TS. It is present in patients with the 45,X karyotype – which is found in about 50–60% of all patients – and in patients with other chromosomal abnormalities.

Prof. Dr. med. M. Ranke
Universitäts-Kinderklinik
Rümelinstrasse 19–23
D–72070 Tübingen (Germany)

0301–0163/95/0449–0035
$8.00/0

Table 1. Adult height in TS in relationship to adult height in normal adult women from various regions

Country	n	Height of normal women, cm	Turner women, cm	Source
Germany	44	167.9±5.2	146.3±6.1	16
Sweden	50	165.9±6.5	146.9±7.8	14
Denmark	76	166.8±5.9	146.8±5.9	17
Belgium	34	163.0±5.5	143.4±5.6	11
USA	536	163.8±6.2	143.3±5.8	18
Italy	105	163.3±8.2	142.5±7.0	19
France	64	163.2±5.6	142.6±5.6	20
UK	138	162.2±6.0	142.9±6.7	15
Japan	18	157.7±4.9	139.5±3.5	21
Europe	661		144.3±6.7	4

Presently there is a wealth of reports on spontaneous development of height [4], height velocity [5, 6] and bone age [5, 7] in TS. These data were partly used to devise disease-specific growth standards. However, the prudent use of these standards requires the following considerations:

(1) TS results from a great variety of chromosomal aberrations which comprise the loss of one X-chromosome (i.e. karyotype 45,X), a structural abnormality of one of the X-chromosomes (e.g. karyotype 46,X i(Xq), and a multitude of combinations of such abnormalities with normal cell lines (e.g. karyotype 46,XX/45,X). Although the 45,X karyotype is prevailing (50–60% in most series), most cohorts contain the less frequent chromosomal abnormalities in varied frequencies. Even though there is presently no convincing evidence that the karyotype has a major impact on growth, it may account for minor variations [8, 9]. In addition, it should be considered that the karyotype identified in peripheral lymphocytes may not accurately reflect that in the tissues relevant for growth.

(2) There is a bias in collection of data through the age when patients are brought to the attention of physicians. For example, the diagnoses may be made before birth through karyotyping or sonographic investigations, the diagnosis may be established through dysmorphology and short stature at various ages, it may be made as a cause of delayed or absent pubertal signs or even infertility. The age at diagnosis may thus be incidental or may be facilitated by the structures within a given health system.

(3) Even within a group of patients with identical karyotypes, the extent of dysmorphology may vary. This is usually not considered when growth data are collected. In particular, there is commonly no consideration of factors like congenital heart failures which may have an impact on growth.

(4) The extent of ovarian failure also varies [10]. Although this appears to have no major influence on final height [11], it has an impact on the tempo of growth probably even before pubertal age.

(5) Data on growth parameters are mostly collected in a mixed cross-sectional/longitudinal manner. The number of observations varies at different ages and is often small at the younger and older ages of a cohort. The sample size on which analyses are based is usually also relatively small when compared to that used to establish standards in normal children.

(6) The data in Turner patients are mostly collected over a long range of years.

(7) Different statistical approaches are taken to formally evaluate the data collected [12].

Thus, in considering all of these aspects, it needs to be kept in mind that from a methodological point of view, all reports fall short to meet the stringent criteria usually applied for establishing auxological standards in the normal population. Nevertheless, despite these limitations, the data on height/length and height velocity (to a lesser degree those on weight and bone age) reported from various sources show a great deal of qualitative homogeneity and have allowed to draw several general conclusions: (1) Patients with TS tend to be small at birth with weight and length being equally affected in magnitude. (2) Height follows the pattern of childhood growth [13, 14] without – or with only minor – superposition of the pubertal growth spurt which therefore has a longer than normal duration. (3) The variance of mean height for a given age is approximately of the same magnitude as in normal children [15, 16] (table 1). (4) There is a high correlation between height (during childhood and in adult life) in TS and parental height [4, 22]. (5) Adult height in TS is on the average 20 cm below the population mean for adult women. (6) The pattern of height development (the shape of the curve of mean height) is similar in cohorts from different countries with populations of similar ethnical background.

A recent survey on spontaneous adult height in several European countries involving 661 patients [4] showed that there are differences in adult height reached depending on the geographical origin: patients from the northern regions tend to be taller than those in the southern regions. These differences are most likely related to the height differences among the underlying populations (table 1) [23], since there is a high overall correlation between target height and adult height. This study again did not show an essential difference between karyotypes. Thus, it is to date rather firmly established how children with TS from different countries and regions grow as a group.

Prediction of Adult Height in TS

In normal children, mean parental height can be used as a rough estimate of adult height. This approach assumes that the genetical influences on adult height from both parents are of an equal magnitude. One common method to calculate the target height of a girl is: (father's height + mother's height):2 minus 6.5 cm. Since in TS adult height is also highly correlated to parental height, a similar approach may be taken [20, 24]. However, as mentioned above, in TS the partial genetical influences of the parents are not always clear. Based on the data from a recent survey [4], the following regression equation for the calculation of the mean prospective height in TS has been derived: adult height [cm] in TS = 75.94 + 0.422 × target height [cm]. This approximation is helpful as a guideline in counseling although the exact figures need further validation.

Methods to predict adult height in normal children are based on the observation that children have reached a certain percentage of adult height at a given bone age. It has been shown that in children with TS the predicted adult height based on the method of Bayley and Pinneau [25] which uses the bone age ratings by the method of Greulich and Pyle [26] is superior in accuracy to other prediction methods [27] and gives a quite good estimate of spontaneous adult height in TS [28, 29]. This is probably due to the fact that after an age of 10 years the percentage of adult height reached in TS is very similar to that in normal children [30]. With the availability of bone age standards specific for TS it may be possible to develop a system for a more accurate prediction of adult height in these children over a wider age range. The most widely used method to predict adult height is the method of 'Projected Adult Height' (PAH). This method is based on the assumption that growth in individuals with TS strictly follows the channel of targeted height [29], and that individual influences on a child's growth (e.g. parental height) are already expressed in the deviation of height from the disease-specific mean for a given age. In practice it is assumed that height expressed in SDS using Turner-specific standards during childhood reflects height SDS in adult life. The validity of this approach for predicting adult height is supported by empirical evidence [15, 29]. In the author's view, this appears to be quite a sound approach. However, it must be strongly emphasized that the results obtained depend on the pattern of the height standards used [6].

It is also assumed that, as in normal children, height for bone age expressed as SDS could be used as a more accurate indicator for adult height. This 'Index of Potential Height' (IPH) was found to be a reasonably accurate means to predict adult height in TS [17, 31]. However, the IPH does not appear to be essentially better than the Bayley-Pinneau method of predicting height in TS which is probably the result of difficulties in bone age rating. Since spontaneous bone age development in TS is different from normal girls, this approach will become more valid when longitudinal standards for bone age in TS will be available. Thus, at present, available methods only allow the prediction of the most likely adult height in TS. An accurate prediction of an individual's adult height with TS is still not possible. However, it should be borne in mind that height predictions in normal children also contain an element of uncertainty.

Evaluation of the Growth Response to Treatment

From a patient's point of view, height in relation to the normal population is of significance. Thus, treatment should attempt to achieve both: growth within normal limits during childhood and adolescence and the attainment of an adult height within the population range. Therefore, in order to evaluate any mode of treatment, the crucial issue is whether the mode improves height compared with disease-specific standards only or whether it really normalizes height. During treatment, an evaluation must be made as to whether or not the mode of treatment applied is in accordance with these two goals. Obviously, an increase in height velocity as such does not help much in assessing the success of treatment – not even if appropriate TS-specific data on spontaneous height velocity are used. The extent of growth is not only determined by height velocity but also by the tempo of growth (e.g. slow growth over a longer period of time may result in higher stature than fast growth over a shorter time). The effect of treatment must, therefore, be evaluated by taking into consideration a parameter which indicates a change in the tempo of growth. During growth of normal children, bone age development is taken as an indicator, since the relationship between height and bone age determines the tempo and the potential of growth. Spontaneous bone age development in TS is different from its normal counterpart [5, 7, 16, 32]. This is partly a result of the absence or very low levels of ovarian steroids. However, in TS other factors may play a role, and the increase in bone age during treatment may differ from the normal at different stages of development. Since bone age ratings are often prone to errors in judgement and are even more difficult to evaluate in TS due to structural abnormalities of the bones, they should be used with caution. Longitudinal data on the natural course of bone age development in TS will be needed before bone age can safely be considered as an indicator when predicting adult stature during the course of spontaneous or medically altered growth in TS.

In practice, the response to growth-promoting therapies is evaluated using different approaches. The simplest approach is using height velocity, either expressed in centimeters/year or in SDS based on Turner-specific standards. As long as patients do not show spontaneous pubertal activity or are treated with sex steroids, this rather simplistic approach has a high descriptive value. A better approach is expressing the response in terms of the changes in height SDS based on Turner-specific standards. The change in height SDS can also be expressed as change in PAH. Both height SDS and PAH do, however, not consider the changes in height potential which may occur as a result of alterations of bone age induced by therapies. Therefore, the development of predicted adult height (see above) may be used as an estimate to document the growth response during therapies. Ultimately, the difference between expected adult height – for example estimated by the method of PAH (see above) – and height actually reached at the end of growth may serve as the best possible parameter to evaluate the response to growth-promoting therapies. Currently, different methods to express the response to growth-promoting therapies in TS are used by various authors, which makes comparison of results difficult and may delay progress of knowledge.

Rationale for Therapy with GH

The reason for short stature in TS is at present not known. Most investigators believe that a disorder at the tissue level is responsible for short stature. Some, however, favor the idea that hormones may partly be a cause. In particular, a disorder of the GH-insulin-like growth factor (IGF) axis has been discussed. There is no doubt that the response of GH to provocative tests is poor, and that the amount of GH secreted spontaneously is frequently subnormal. Particularly at pubertal age, GH is definitely below the levels found in normal girls during puberty [33–35]. Likewise, low or subnormal IGF-I levels have been found during that period [34, 36]. This, however, has been attributed to the absence of estrogen and mild obesity, since the conditions can be reversed by estrogens and/or reduction in weight [37]. The serum levels of the GH-dependent IGF-binding protein (IGFBP-3), which is highly correlated with spontaneous GH secretion, were found to be within the normal range [38]. Thus, there is no convincing evidence that a GH deficiency (GHD) is involved in the pathogenesis of the growth disorder in TS.

Treatment with GH

When pituitary human GH became available, a few patients with TS were treated with it [39–41]. The results of these trials were largely disappointing. With the recent availability of GH produced by the recombinant technique in abundant amounts, it has been possible to conduct carefully designed studies with the aim of re-evaluating this method [42–46]. In contrast to the observations made when pituitary human GH was used, the growth rates in recent studies using recombinant GH were, without exception, reported to be higher than those spontaneously observed in TS. The causes for the observed changes in the results with GH are a matter of speculation. Part of the reason is that the total dosages applied at present are higher. In addition to this, the bioavailable GH dose is increased by giving more frequent injections. Although the response depends on the GH dose, it is, nevertheless, much lower than that observed with true GHD. In combination with oxandrolone, considerably higher growth rates have been observed [44, 47, 48]. This is probably a result of the different mechanisms of action of the two drugs. The number of reports on the long-term results of treatment with GH, in particular reports on final height, in patients with TS is still very limited. The most extensive data reported are those of Rosenfeld et al. [49]. Of the originally 65 patients who entered the study at a mean age of 9.3 years, 30 have discontinued therapy after a mean duration of treatment of 54 months and at a mean age of 16.7 years. The patients received a weekly dose of GH of approximately 1.0 IU/kg BW (0.375 mg/kg BW), sometimes combined with oxandrolone (0.125–0.065 mg/kg BW/day). The gain above projected adult height averaged 8.1 cm. The results of several other studies on treat-

Table 2. Long-term results of GH treatment in Turner syndrome

Author	Ref.	Year	n	Age	HT SDS[1]	GH dose	Injections/ week	Years on GH	dHT SDS[2]	dPAH cm
Job	59	1991	10	10.1	−3.1	0.7	7		3	2.1
			12	11.4	−3.7	0.7	7		3	4.5
Lenko	54	1993	25	11.6	−3.4	0.7	7	2	1.0	3.1
Knudtzon	53	1993	33	11.8	−3.2	0.7	7	2	1.15	5.5
				14		0.7[3]	7	4	2.0	
Nilsson	60	1993	14	12.1		1.0[4]	7	4	1.8	
Takano	50	1993	23	10.8		0.5	6/7	5	1.0	
			36	10.1		1.0	6/7	5	1.5	
Nienhuis	52	1993	9			0.8	3	4	1.3	2.8
			10			0.8	6	4	1.7	7.8
			10			0.8[3]	6	4	1.1	9.8
Ranke	51	1994	137	10.5	−3.1	0.8	6/7	3	1.3	
			14	10.3	−2.7	0.8[4]	6/7	3	1.2	
			47	12.0	−3.2	0.8[3]	6/7	3	0.9	

1 SDS based on normal children.
2 SDS based on Turner standards.
3 GH plus estrogen.
4 GH plus oxandrolone.
d = Delta; GH dose = IU/kg BW/week; PAH = predicted adult height.

ment for 2 or more years are listed in table 2. Since the modalities of evaluating the response to treatment vary, the list gives only a limited view of the work done in this field. Nevertheless, the results show a common pattern. It can be seen that patients show an improvement of height when treated over many years. As has been shown by Takano et al. [50] and Ranke [51], the magnitude of the long-term gain is a function of the GH dose. As shown by Nienhuis et al. [52], the frequency of GH administrations over the week is positively correlated with the magnitude of the response – as it is in patients with GHD. In the studies, the patient characteristics with respect to mean age at onset of therapy and the modes of therapy are in general quite similar, and the observed long-term gain in height does not vary a great deal. From the data one can assume that most of the benefit occurs during the first 2 years of treatment. One possible explanation for this could be that the age at onset of therapy is already so high that after 2 years the potential to gain height through treatment is dramatically reduced. The weaning of the GH effect may be another explanation. Since the long-term effect on growth was found to be negatively correlated with the age of onset of treatment [51], earlier treatment is suggestive. There is an ongoing discussion about whether or not GH treatment may inappropriately advance bone age development – perhaps in an age-dependent manner. This issue is not settled by the analyses presently available. While, for example, Knudtzon et al. [53] were unable to show an advancement of bone age development, the observations of Lenko et al. [54] support the opposing view. The fact that the changes in delta height SDS are in the same order of magnitude as those in PAH is, however, in line with the assumption that GH itself, in the long run, does not advance bone age unduly. This is probably not so if GH is combined with estrogen and/or oxandrolone in Turner children before pubertal age.

At this stage it seems fair to state that GH (alone or in combination with oxandrolone) is not only able to increase the tempo of growth but also adult height. However, it is not at all clear which treatment modalities need to be chosen to reach the goals of height normalization during childhood and in adult life.

Risks of Growth-Promoting Therapies

The potential risks of growth-promoting therapies can be brought about by the following: (1) specific factors related to TS; (2) specific side effects of the drugs used; (3) the dosage and duration of the drugs used, and (4) a combination of these factors. At present, we know rather little about the spontaneous long-term risks of patients with TS, since there are no data available on TS in adult life [55]. Impaired glucose tolerance and a tendency towards high blood pressure (even without cardiac failure) are well-documented abnormalities in TS [for review, see 56, 57]. High doses of GH may increase the risk of diabetes mellitus developing. Although recent studies have shown that the negative effect of GH on glucose tolerance was transient [48, 56] and that the incidence of diabetes mellitus in Turner patients on GH is not higher than in the normal population [58], the long-term risks cannot be predicted. It is also noteworthy, although on the basis of the underlying symptomatology in TS probably not remarkable, that a higher than normal incidence of edema and kyphoscoliosis was reported during GH treatment in TS [58]. The risk potential of estrogens and anabolic steroids is even more difficult to assess. Oxandrolone has been in use for decades and appears to be rather safe when given in low doses. However, its potential hazards when given over long periods of time are by no means established. Since the dose-risk potential may be age-dependent, care must be taken when treating younger patients.

References

1 Turner HH: A syndrome of infantilism, congenital webbed neck and cubitus valgus. Endocrinology 1938;23:566–574.
2 Ullrich O: Über typische Kombinationsbilder multipler Abartung. Z Kinderheilkd 1930;49:271–276.
3 Wiedemann HR, Glatzl J: Follow-up of Ullrich's original patient with 'Ullrich-Turner' syndrome. Am J Med Genet 1991;41:134–136.
4 Ranke MB, Grauer ML: Adult height in Turner syndrome: Results of a multinational survey. Horm Res 1994;42:90–94.
5 Ranke MB, Chavez-Meyer H, Blank B, Frisch H, Häusler G: Spontaneous growth and bone age development in Turner syndrome – Results of a multicentric study 1990; in Ranke MB, Rosenfeld RG (eds): Turner Syndrome: Growth-Promoting Therapies. Amsterdam, Excerpta Medica/Elsevier, 1991, pp 101–106.
6 Ranke MB: Growth in Turner's syndrome. Acta Paediatr 1994;83:343–344.
7 Rochiccioli P, Pienkowski C, Tauber MT: Spontaneous growth in Turner syndrome: A study of 61 cases; in Ranke MB, Rosenfeld RG (eds): Turner Syndrome: Growth-Promoting Therapies. Amsterdam, Excerpta Medica/Elsevier, 1991, pp 107–112.
8 Mazzanti L, Tassinari D, Bergamaschi R, Magnani C, Chiumello G, Cacciari E: Spontaneous growth and pubertal development in Turner's syndrome with different karyotypes. Acta Paediatr 1994;83:299–304.
9 Hall J: The relationship between karyotype and growth in Turner syndrome; in Ranke MB, Rosenfeld RG (eds): Turner Syndrome: Growth-Promoting Therapies. Amsterdam, Excerpta Medica/Elsevier, 1991, pp 9–13.
10 Hibi I, Tanae A, Tanaka T, Yoshizawa A, Miki Y, Ito J: Spontaneous puberty in Turner syndrome: Its incidence, influence on final height and endocrine features; in Ranke MB, Rosenfeld RG (eds): Turner Syndrome: Growth-Promoting Therapies. Amsterdam, Excerpta Medica/Elsevier, 1991, pp 75–81.
11 Massa G, Vanderscheuren-Lodeweyckx M, Craen M, Vanderweghe M, van Vliet G: Growth hormone treatment of Turner syndrome patients with insufficient growth hormone response to pharmacological stimulation tests. Eur J Pediatr 1991;150:460–463.
12 Häusler G, Frisch H: Methods for evaluation of growth in Turner's syndrome: Critical approach and review of the literature. Acta Paediatr 1994;83:309–314.
13 Karlberg J, Albertsson-Wikland K, Naeraa RW: The Infancy-Childhood-Puberty (ICP) model of growth for Turner girls; in Ranke MB, Rosenfeld RG (eds): Turner Syndrome: Growth-Promoting Therapies. Amsterdam, Excerpta Medica/Elsevier, 1991, pp 89–94.
14 Karlberg J, Albertsson-Wikland K, Naeraa RW, Rongen-Westerlaken C, Wit JM, in collaboration with the Swedish and the Dutch Study Group for GH Treatment: Reference values for spontaneous growth in Turner girls and its use in estimating treatment effects; in Hibi I, Takano K (eds): Basic and Clinical Approach to Turner Syndrome. Amsterdam, Excerpta Medica, 1993, pp 83–92.
15 Lyon AJ, Preece MA, Grant DB: Growth curve for girls with Turner's syndrome. Arch Dis Child 1985;60:932–935.
16 Ranke MB, Pflueger H, Rosendahl W, Stubbe P, Enders H, Bierich JR, Majewski F: Turner syndrome: Spontaneous growth in 150 cases and review of the literature. Eur J Pediatr 1983;141:81–88.
17 Naeraa RW, Nielsen J: Standards for growth and final height in Turner's syndrome. Acta Paediatr Scand 1990;79:182–190.
18 Lippe B, Frane J: Growth in Turner syndrome: The United States experience; in Ranke MB, Rosenfeld RG (eds): Turner Syndrome: Growth-Promoting Therapies. Amsterdam, Excerpta Medica/Elsevier, 1991, pp 59–65.
19 Bernasconi S, Larizza D, Benso L, Volta C, et al: Turner's syndrome in Italy: Familial characteristics, neonatal data, standards for birth weight and for height and weight from infancy to adulthood. Acta Paediatr 1994;83:292–298.
20 Salerno MC, Job JC: La taille dans le syndrome de Turner: Corrélation avec la taille des parents. Arch Fr Pédiatr 1987;44:863–865.
21 Suwa S: Standards for growth and growth velocity in Turner's syndrome. Acta Paediatr Jpn 1992;34:206–221.
22 Brook CG, Gasser T, Werder EA, Zachmann M, Prader A: Height correlations between parents and mature offspring in normal subjects and in subjects with Turner's and Klinefelter's and other syndromes. Ann Hum Biol 1977;4:17–22.
23 Rochiccioli P, David M, Malpuech G, Colle M, Limal JM, Battin J, Mariani R, Sultan C, Nivelon JL, Simonin G, Bost M, Chatelain P, Bertrand AM, Freycon R, Bouquier JJ, Pierron P: Study of final height in Turner's syndrome: Ethnic and genetic influences. Acta Paediatr 1994;83:305–308.
24 Lenko HL, Perheentuppa J, Soederholm A: Growth in Turner's syndrome: Spontaneous and fluoxymestrone stimulated. Acta Paediatr Scand Suppl 1979;277:57–63.
25 Bayley N, Pinneau S: Tables for predicting adult height from skeletal age. J Pediatr 1952;40:423–441.

26 Greulich W, Pyle S: Radiographic Atlas of Skeletal Development of the Hand and Wrist, ed 2. Stanford, Stanford University Press, 1959.
27 Zachmann M, Sobradillo B, Frank M, Frisch H, Prader A: Bayley-Pinneau, Roche-Wainer-Thissen, and Tanner height predictions in normal children and in patients with various pathological conditions. J Pediatr 1978;93: 749–755.
28 Naeraa RW, Eiken M, Legarth EG, Nielsen J: Spontaneous growth, final height and prediction of final height in Turner syndrome; in Ranke MB, Rosenfeld RG (eds): Turner Syndrome: Growth-Promoting Therapies. Amsterdam, Excerpta Medica/Elsevier, 1991, pp 113–116.
29 Frane JW, Sherman BW, and the Genentech Collaborative Group: Predicted adult height in Turner syndrome; in Rosenfeld RG, Grumbach MM (eds): Turner Syndrome. New York, Dekker, 1990, pp 393–403.
30 Ranke MB, Bierich JR, Stubbe P, Majewski F: Spontaneous growth in Ullrich-Turner syndrome; in Rosenfeld RG, Grumbach MM (eds): Turner Syndrome. New York, Dekker, 1990, pp 149–162.
31 Joss E: Evaluation of hormonal treatment on linear growth and skeletal maturation: Methods of predicting final height in Turner syndrome; in Ranke MB, Rosenfeld RG (eds): Turner Syndrome: Growth-Promoting Therapies. Amsterdam, Excerpta Medica/Elsevier, 1991, pp 83–88.
32 Brook CGD, Mürset G, Zachmann M, Prader A: Growth in children with 45,XO Turner's syndrome. Arch Dis Child 1974;49:789–795.
33 Buchanan CR, Law CM, Milner RDG: Growth hormone in short, slowly growing children and those with Turner's syndrome. Arch Dis Child 1987;62:912–916.
34 Ranke MB, Blum WF, Haug F, Rosendahl W, Attanasio A, Enders H, Gupta D, Bierich JR: Growth hormone, somatomedin levels and growth regulation in Turner's syndrome. Acta Endocrinol (Copenh) 1987;116:305–313.
35 Ross JL, Long LM, Loriaux LD, Cutler GB: Growth hormone secretory dynamics in Turner syndrome. J Pediatr 1985;106:202–206.
36 Cuttler L, Van Vliet G, Conte FA, Kaplan SL, Grumbach MM: Somatomedin-C levels in children and adolescents with gonadal dysgenesis: Differences from age-matched normal females and effect of chronic estrogen replacement therapy. J Clin Endocrinol Metab 1985;60: 1087–1092.
37 Schober E, Frisch H, Waldhauser F, Biegelmayer CH: Influence of estrogen administration on growth hormone stimulation in patients with Turner's syndrome. Acta Endocrinol (Copenh) 1987;283(suppl):175–176.
38 Ranke MB, Blum WF, Frisch H: The acid-stable subunit of insulin-like growth factor binding protein in disorders of growth; in Drop SLS, Hintz RL (eds): Insulin-Like Growth Factor Binding Proteins. Amsterdam, Excerpta Medica/Elsevier, 1989, pp 103–112.
39 Hutchings J, Escamilla R, Li C, Forsham P: Li human growth hormone administration in gonadal dysgenesis. Am J Dis Child 1965;109: 318–321.
40 Tanner J, Whitehouse R, Hughes P, Vinve F: Effect of human growth hormone treatment for 1 to 7 years on growth of 100 children with growth hormone deficiency, low birth weight, inherited smallness, Turner's syndrome and other complaints. Arch Dis Child 1971;46: 745–782.
41 Tzagournis M: Response to long-term administration of human growth hormone in Turner's syndrome. J Am Med Assoc 1969;210:2373–2376.
42 Raiti S, Moore WV, Van Vliet G, Kaplan SL, and the National Hormone and Pituitary Programme: Growth-stimulating effects of human growth hormone therapy in patients with Turner syndrome. J Pediatr 1986;109:944–949.
43 Rongen-Westerlaken C, Wit JM, Drop SLS, Otten BJ, Oosdijk W, Delemarre-Van De Waal HA, Gons MH, Bot A, Van den Brande JL: Methionyl human growth hormone in Turner's syndrome. Arch Dis Child 1988;63:1211–1217.
44 Rosenfeld RG, Hintz RL, Johanson AJ, Brasel JA, Burstein S, Chernausek SD, Clabots T, Frane J, Gotlin RW, Kuntze J, Lippe BM, Mahoney PC, Moore WV, New MI, Saenger P, Stoner E, Sybert V: Methionyl human growth hormone and oxandrolone in Turner's syndrome. Preliminary results of a prospective randomized trial. J Pediatr 1986;109:936–943.
45 Singer-Granick C, Lee PA, Foley TP, Becker DJ: Growth hormone therapy for patients with Turner's syndrome. Horm Res 1986;24:246–250.
46 Takano K, Hizuka N, Shizume K: Growth hormone treatment in Turner's syndrome. Acta Paediatr Scand Suppl 1986;325:58–63.
47 Nilsson KO, Wattsgard C, Albertsson-Wikland K, et al: Growth-promoting treatment in Turner syndrome – Three-year experience in Sweden, preliminary results; in Ranke MB, Rosenfeld RG (eds): Turner Syndrome: Growth-Promoting Therapies. Amsterdam, Excerpta Medica/Elsevier, 1991, pp 215–219.
48 Stahnke N, Stubbe P, Keller E, Zeisel HJ, and the Serono Study Group Hamburg: Effects and side-effects of GH plus oxandrolone in Turner syndrome; in Ranke MB, Rosenfeld RG (eds): Turner Syndrome: Growth-Promoting Therapies. Amsterdam, Excerpta Medica/Elsevier, 1991, pp 241–248.
49 Rosenfeld RG, Frane J, Attie KM, Burstein S, Cara JF, Charnausek S, Gothlin RW, Kuntze G, Lippe BM, Mahoney PC, Moore WV, Saenger P, Johanson AJ: Six-year results of a randomized, prospective trial of human growth hormone and oxandrolone in Turner syndrome. J Pediatr 1992;121:49–55.
50 Takano K, Shizume K, Hibi I, Ogawa Y, Suwa S, Tanaka T: Growth hormone treatment in Turner syndrome: Results of a multicenter study in Japan. Horm Res 1993;39(suppl):37–41.
51 Ranke MB: Turner syndrome: Demography, auxology and growth during growth hormone treatment in the Kabi International Growth Study; in Ranke MB, Gunnarsson R (eds): Progress in Growth Hormone Therapy – 5 Years of KIGS. Mannheim, J&J Verlag, 1994, pp 190–205.
52 Nienhuis HE, Rongen-Westerlaken C, Wit JM, Otten BJ, de Muinck Keizer-Schrama SMPF, Drayer NM, Delemarre-van de Waal HA, Vulsma T, Oostdijk W, Waelkens JJJ: Results of long-term therapy with growth hormone in two dose regimes in Turner syndrome. Horm Res 1993;39(suppl):31–36.
53 Knudtzon J, Aaarskog D: Results of two years of growth hormone treatment followed by combined growth hormone and oestradiol in Turner syndrome. Horm Res 1993;39(suppl):7–17.
54 Lenko HL, Hakulinen A, Käär M-L, Mäkelä A-L, Sipilä I: Effect of conventional dose of growth hormone therapy for two years on height velocity and height prognosis in girls with Turner syndrome. Horm Res 1993; 39(suppl):3–6.
55 Price DA, Clayton PE, Crowne EH, Roberts CR: Safety and efficacy of human growth hormone treatment in girls with Turner syndrome. Horm Res 1993;39(suppl):44–48.
56 Wilson DM, Frane J, Sherman B, et al: Carbohydrate and lipid metabolism in Turner syndrome: Effect of therapy with growth hormone, oxandrolone, and a combination of both. J Pediatr 1988;112:210–217.
57 Chiumello G, Bognetti E, Bonfanti R, Ciralli F, Cofano D: Glucose metabolism in Turner syndrome; in Ranke MB, Rosenfeld RG (eds): Turner Syndrome: Growth-Promoting Therapies. Amsterdam, Excerpta Medica/Elsevier, 1991, pp 47–49.
58 Wilton P: Adverse events during growth hormone treatment. Five years' experience in the Kabi International Growth Study; in Ranke MB, Gunnarsson R (eds): Progress in Growth Hormone Therapy – 5 Years of KIGS. Mannheim, J&J Verlag, 1994, pp 291–307.
59 Job JC, Landie F: Three-year results of treatment with growth hormone, alone or associated with oxandrolone, in girls with Turner syndrome. Horm Res 1991;35:229–233.
60 Nilsson KO, Albertsson-Wikland K, Aronson S, Gustafsson J, Hagenäs L, Häger A, Ivarsson SA, Karlberg J, Kriström B, Moell C, Ritzén M, Tuvemo T, Wattsgard C, Westgren U, Westphal O, Aman J: Long-term results of combination treatment with growth hormone, oxandrolone and ethinyl estradiol in girls with Turner syndrome; in Hibi I, Takano K (eds): Basic and Clinical Approach to Turner Syndrome. Amsterdam, Excerpta Medica/Elsevier, 1993, pp 303–310.

Horm Res 1995;44(suppl 3):42–48

Cinzia Galasso
Giuseppe Scirè
Brunetto Boscherini

Department of Pediatrics,
University 'Tor Vergata', Rome, Italy

Growth Hormone and Dysmorphic Syndromes

Key Words
Short stature
Growth hormone
Dysmorphic syndrome

Abstract
In order to study the pathogenesis of short stature in some of the best known and most frequent dysmorphic syndromes, we have reviewed the most significant studies conducted on somatrotropin secretion and on response to replacement treatment with human growth hormone in pediatric patients. In particular, the study examines the results presented in the literature, and in a few of our cases, those obtained with regard to Noonan, Silver-Russell and Prader-Willi syndrome patients, to achondroplasia and hypochondroplasia patients, and to Down syndrome patients. Finally, we shall present a review of a few, less frequent dysmorphic syndromes with short stature, in which a growth hormone deficiency has been diagnosed and replacement treatment attempted.

Introduction

The possible treatment of short stature in dysmorphic syndromes has aroused great interest in the past few years. There are many reasons for such interest. First of all, some authors [1] have observed that the therapy proved effective in patients with Turner syndrome, a condition in which short stature is not secondary to growth hormone (GH) deficiency in the classical sense. Also, the wide availability of synthetic GH allowed researchers to experiment with this treatment in various dysmorphic syndromes, frequently obtaining good results in a short period. Finally, some of the data contained in the literature leads to think that a GH deficiency may be present, even if exceptionally, in some patients with short stature and dysmorphic features [2].

In this paper we will present the data contained in the literature concerning the study of GH function in the most important dysmorphic syndromes, as well as the results of GH treatment in a few patients. We will also describe (table 1) a few, less frequent syndromes in which a GH deficiency has been described, if only occasionally.

Noonan Syndrome

The study of somatotropin function in patients with Noonan syndrome has yielded conflicting results. Values obtained both by pharmacological stimulation and from the study of spontaneous secretion are sometimes normal and other times insufficient [3–7]. Serum insulin-like growth factor 1 (IGF-1) was also found in a few subnormal patients [5, 8]. However, in the experience of many authors, the results produced by GH treatment were not easy to interpret. Ahmed et al. [5] obtained a good response to GH therapy at the dosage of 18 U/m^2 s.c. in 3

Dott. Cinzia Galasso
via Giuseppe Pianese 9
I-00151 Roma (Italy)

0301–0163/95/0449–0042
$8.00/0

Table 1. Dysmorphic syndromes: GH secretion and treatment

Syndrome	Study first author	ref.	Cases with GH deficit	Cases treated	Time of treatment years	Results of short-term therapy (<1 year)	Results of long-term therapy (>1 year)
Donahue	Togoe	52	1	1	2	+	+
Poland	Larizza	53	1	1	1	+	–
Johanson Blizzard	Kristjansson	54	1	–	–	–	–
Cohen	Massa	55	1	1	2.5	+	+
Hallermann-Streiff	Pivnick	56	1	–	–	–	–
Cornelia de Lange	Schwartz	57	2	–	–	–	–
Esostosi Multipla	Martinez Boscherini	58 70	1 1	1 1	1 15	+ +	– +
Aarskog	Kodama	59	1	–	–	–	–
Bloom	Vanderschueren-Lodeweyckx	60	1	–	–	–	–
Lery-Weill	Zucchini Thuestad	61 62	Borderline 5	1 5	2 1 (5 patients) 2 (3 patients)	+ + (5 patients)	+ + (3 patients)
CHH	Lenko	63	–	2	2	neg	neg
Ring Chromosome 18	Abusrewil Meloni	64 65	1 1	– 1	– 1	– +	– –
LEOPARD	Malpuech	66	1	–	–	–	–
Fanconi	Pochedly Zachmann Gleadhill	67 68 69	1 1 1	– 1 1	– 2.2 1	– + neg	– + neg
Abnormal face ...	Boscherini	70	1	1	9	+	+
Mental retardation ...	Chitayat	71	2	1	4	+	+
Growth hormone ...	Stratton	72	1	–	–	–	–
Kabuki make-up	Togoe	52	1	1	4	+	+
Mulibrey	Haraldsson Lenko	73 63	1 2	1 5	2 2 (5 patients) 3 (4 patients) 4 (1 patient)	+ +	+ +

+ = Positive result; neg = negative result.

of the 5 patients treated for 1 year. Recently, Thomas and Stanhope [6] used supraphysiological doses (20–30 U/m² s.c./week) in 5 patients, only 1 of whom had low GH secretion, and obtained a stature increase after 1.8–4 years of treatment. The average stature grew from –3.3 to –2.4 SD, with a satisfactory growth velocity increase in the first year. In the following years, however, growth velocity went back to pretreatment values.

Our experience involved 7 children. In 3 of them, therapy proved ineffective and was discontinued after 6–9 months. However, since a low GH dosage (9 U/m² s.c./week) was used in those first 3 patients, treatment was

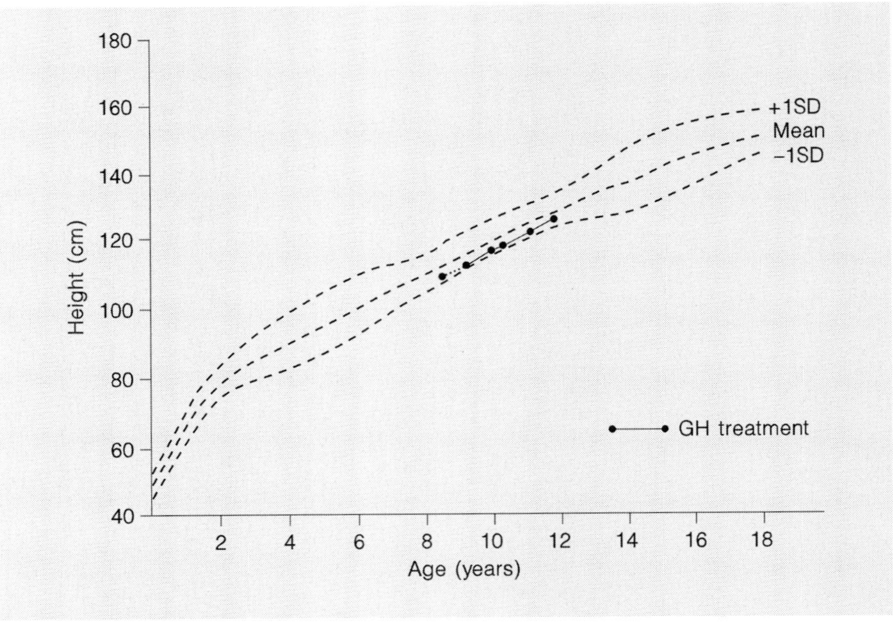

Fig. 1. Growth curve of a girl with Noonan syndrome treated with GH.

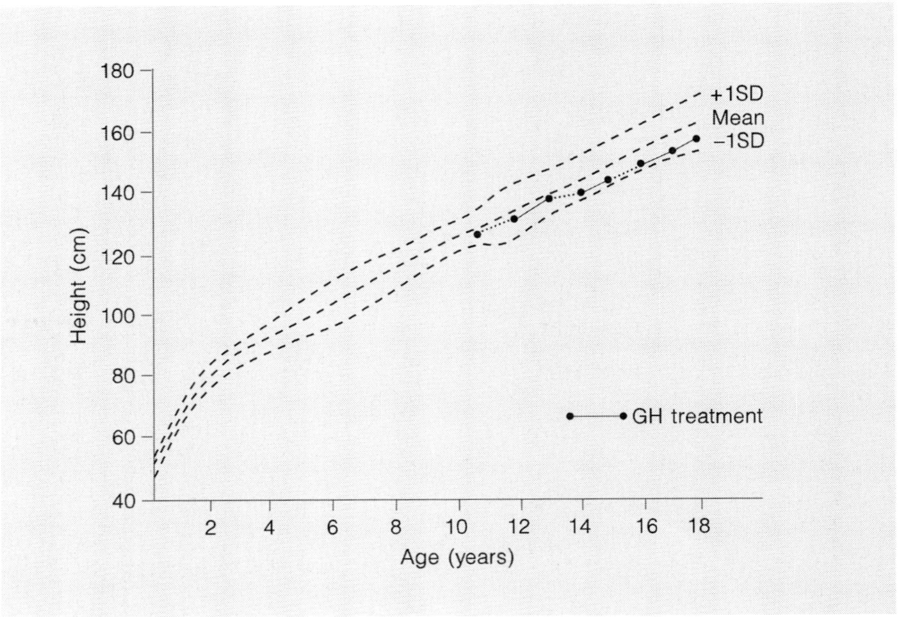

Fig. 2. Growth curve of a boy with Noonan syndrome treated with GH.

tried in 2 other patients with a higher dosage (12 U/m^2 s.c./week). During the first year of treatment, we obtained a satisfactory increase of growth velocity. Also, 2 other patients received the treatment at the same dosage for a few years, and in those we were able to observe the final stature. In spite of a satisfactory initial increase in growth velocity, growth retardation in adult patients was almost the same as at the beginning of treatment (fig. 1, 2).

To conclude, GH treatment in subjects with Noonan syndrome is certainly effective in the first year, while long-term results are yet to be evaluated. In any event, supraphysiological GH dosages should be used in long-term treatment. Before beginning treatment in these patients, it should however be emphasized that short stature is frequently observed in children [9–11], but that only about 50% of adult patients will show a stature below the third centile of the normal population [12, 13].

Silver-Russell Syndrome

Since somatotropin secretion in patients with Silver-Russell syndrome is usually normal [14], the cause of short stature is thought to be secondary to intrauterine growth retardation. However, GH deficiency was documented in a few patients [14–20]. The deficiency was isolated in 3 patients [14–16, 19]; in the others, it was associated with deficiency of other pituitary hormones [18, 20, 21]. Also, 1 patient with Silver-Russell syndrome was described in whom panhypopituitarism was secondary to a craniopharyngioma [22].

Recently, Stanhope et al. [23] studied the spontaneous nocturnal GH secretion in 14 subjects with IUGR and in 17 patients with Silver-Russell syndrome. From the data presented, GH insufficiency appears to be common in children with short stature secondary to IUGR; in patients with Silver-Russell syndrome, a relative GH insufficiency may be more frequent and the anomaly of the nocturnal GH secretion pattern observed (a single nocturnal GH pulse) may be an additional reason for abnormal growth in these patients.

GH treatment for 1 year usually determines a growth velocity increase [14, 16, 20, 21, 24]; if therapy is extended over 1 year, instead, results are contradictory. Tanner et al. [14] treated 19 patients for 1 year obtaining no results. Chatelain [25] instead, in a study of 45 children with noncontrolled diagnosis, obtained good results in patients treated for 3 years, whose stature grew from –3.5 to –2.2 SD. Unfortunately, end-of-treatment bone age is not reported in this work; therefore, the effective gain in terms of final stature is not known. In fact, an acceleration of bone maturity was already observed in these patients during treatment [20, 23, 24, 26].

Ultimately, GH treatment produces an increase in growth velocity in the first year of therapy, but an effective long-term improvement is far from being demonstrated in patients with normal pituitary function. Since GH deficiency, even if only as an exception, may be present in children with Silver-Russell syndrome, it is advisable to conduct a study of GH secretion in view of a possible replacement treatment. If a GH deficit is present, however, an MRI should be performed in order to exclude with certainty a possible panhypopituitarism [17, 18, 20, 21]. In our experience, one baby with panhypopituitarism with hypoplasia documented by MRI presented a phenotype identical to Silver-Russell syndrome. Another child with Silver-Russell syndrome and GH deficiency was treated with GH for 12 years. The final height (SD) was the same as at the beginning of treatment (fig. 3).

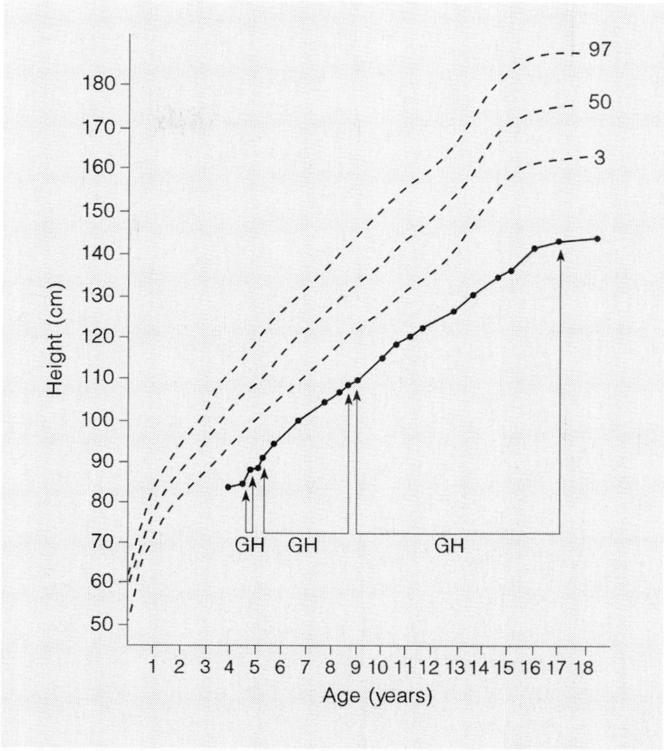

Fig. 3. Growth curve of a boy with Silver-Russell syndrome treated with GH.

Prader-Willi Syndrome

Short stature is one of the syndrome's distinctive features [27]. Its cause has not yet been clearly established, although the involvement of a hypothalamic-pituitary dysfunction has been suspected for years [28]. Low GH concentrations were found in a few patients [29, 30], but the interpretation of that data was made difficult by the almost invariable presence of obesity, a condition in which reduced response to pharmacological stimulation is frequently observed [31].

Studies on somatotropin secretion were resumed since a few authors [32] found abnormal GH secretion in 5 non-obese patients. Based on their data, those authors believed that short stature in subjects with Prader-Willi syndrome was caused by an effective GH deficiency and was not a consequence of obesity. Reduced GH secretion was also observed by other authors [33], who used a more powerful stimulation than those traditionally used [34]. A true GH deficiency was present in two patients.

Early experiments with GH therapy were made many years ago [35] and the first results, however obtained using low GH dosages (6 U/week), were negative. More recently, GH treatment was attempted again [28]. In 2 patients with normal GH secretion and low levels of IGF-1, the treatment conducted for over 1 year produced results which were considered as positive by the authors. In particular, no acceleration of bone maturity or other negative side effects were observed. In 1 of the 2 patients, the addition of oxandrolone determined a further increase in growth velocity.

These promising results will have to be confirmed, and it will be necessary to rule out with certainty that long-term treatment causes sugar intolerance, a condition for which these patients are predisposed, or other unwanted side effects [36, 37].

Achondroplasia – Hypochondroplasia

The pathogenic connection between achondroplasia and hypochondroplasia has not yet been established: some authors view them as distinct, others consider the latter as an attenuated form of the former [38–40]. The GH status in achondroplasia is incompletely clarified, but normal GH response to standardized *L*-dopa and arginine tests have been reported, and normal plasma levels of IGF-1 have been found [37]. Normal GH values after pharmacological stimulation [41, 42] and during sleep [41] have been observed again more recently.

Somatotropin secretion was also studied in hypochondroplastic children: Appan et al. [43] demonstrated normal GH responses to a pharmacological test in 20 subjects. Kodama et al. [41] report a modest increase in growth velocity (DHV = +2 cm/year) after 2 years of GH treatment at the dosage of 0.5 U/kg/week in 1 child with achondroplasia. Okabe et al. [42] treated 3 children with achondroplasia and 1 with hypochondroplasia, obtaining a good response in terms of growth velocity, but for an even shorter period (6–12 months). In both works, there is no indication of an acceleration of bone age.

Horton et al. [44] studied the effects of GH treatment in 6 children with achondroplasia at the dosage of 0.3 mg/kg/week for 6 months. GH therapy modestly increased growth velocity only in part of the children.

Observations on children with hypochondroplasia, although conducted over longer periods of time, do not provide clear indications on the actual usefulness of GH treatment. Studies by Appan et al. [43] and by Bridges et al. [45] indicated a positive response after the first year of treatment in terms of growth velocity, but an inadequate response in the 2 following years.

Down Syndrome

We know that children with Down syndrome show a mild growth retardation at birth. Growth velocity is normal until 6 months, but is then reduced between that age and 3 years. Torrado et al. [46] observed GH deficiency after pharmacological stimulation or spontaneous 24-hour GH secretion in some of these patients. Furthermore, Anneren et al. [47] demonstrated normal GH secretion after pharmacological tests, but low radioimmunoassays for serum levels of IGF-1. They suggested that Down syndrome might involve a delayed maturation of the IGF system. These results are in accordance with those of Ragusa et al. [48]. Based on those results, Torrado et al. [46] treated 13 children with Down syndrome and microcephaly for 1 year with recombinant GH at the dosage of 0.3 mg/kg/week, obtaining an increase of growth velocity from 5.4 to 12.2 cm/year. Good results were also obtained by Anneren et al. [49] in treating 16 patients for 12 months, 12 for 24 months, and 8 for 30 months, with no acceleration of bone age after 1 year of therapy. However, many doubts remain [50] concerning the effectiveness and possible risks of this treatment.

The National Down Syndrome Society in the United Kingdom and the Science Advisory Board of the National Syndrome Society in the USA [51] recommend GH treament in children with Down syndrome exclusively within adequately controlled studies. Otherwise, it should not be used.

Therefore, we believe the further considerations should be made with regard to GH therapy in subjects with Down syndrome: (1) Short-term GH response is not sufficient to justify the treatment of all children with Down synrome. A careful study of the growth velocity and of the bone maturation pattern even in the years following treatment is essential to assess the usefulness of a treatment which might produce temporary pharmacological effects, but no true gains in terms of final stature. (2) We do not believe that the benefit 'which is only hypothetical' of a slight stature increase can warrant the considerable psychological strain of such a complex treatment in mentally retarded children. (3) Furthermore, doubts remain that the risk of leukemia, already higher in these subjects, may further increase during the GH therapy.

Therefore, we believe that the benefits of GH therapy in children with Down syndrome should be carefully weighed against the risks implied, whether certain or only hypothetical.

Conclusion

To conclude, GH treatment of short stature in subjects with dysmorphic syndromes is often very effective in increasing the growth velocity in the first year of treatment. However, when treatment is continued for a longer time, results become uncertain or almost always negative, especially since GH treatment induces an acceleration in bone maturity that nullifies the apparent improvement in terms of stature. Another conclusion that can be drawn from the literature is that the presence of short stature in dysmorphic syndromes does not necessarily rule out a GH deficiency; therefore, a study of GH function should always be conducted when growth data seem to indicate a GH deficiency.

References

1 Rosenfeld RG, Frane J, Attie KP, Brasel JA, Burstein S, Cara JF, Chernausek S, Gotlin RW, Kuntze J, Lippe BM, Mahoney PC, Moore WV, Saenger P, Johanson AJ: Six-year results of a randomized, prospective trial of human growth hormone and oxandrololone in Turner syndrome. J Pediatr 1992;121:49–55.

2 Boscherini B, Galasso C, Carnazza S: Genetic Syndromes and GH Deficit. New York, Raven Press, 1990, pp 199–125.

3 Cianfarani S, Spadoni GL, Finocchi G, Ravet P, Costa F, Papa M, Scirè G, Manca Bitti ML, Boscherini B: Trattamento con ormone della crescita in tre casi di sindrome di Noonan. Minerva Pediatr 1987;39:281–284.

4 Ross GL, Shenkman L: Noonan's syndrome and hypopituitarism. Am J Med Sci 1980;279: 47–52.

5 Ahmed ML, Foot ABM, Edge JA, Lamkin VA, Savage MO, Dunger DB: Noonan's syndrome: Abnormalities of the growth hormone/IGF-1 axis and the response to treatment with human biosynthetic growth hormone. Acta Paediatr Scand 1991;80:446–450.

6 Thomas BC, Stanhope R: Long-term treatment with growth hormone in Noonan's syndrome. Acta Paediatr 1993;82:853–855.

7 Cianfarani S, Vaccaro F, Boscherini B: What is the rationale for growth hormone therapy in Turner's syndrome? Lancet 1994;344:114–115.

8 Bernardini S, Spadoni GL, Cianfarani S, Vaccaro F, Pasquino AM, Boscherini B: Growth hormone secretion in Noonan's syndrome. J Pediatr Endocrinol 1991;4:217–221.

9 Collins E, Turner G: The Noonan syndrome – A review of the clinical and genetic features of 27 cases. J Pediatr 1973;83:941–950.

10 Nora JJ, Nora AH, Sinha AK, Spangler RD, Lubs HA: The Ullrich-Noonan syndrome (Turner phenotype). Am J Dis Child 1974;127: 48–55.

11 Sharland M, Burch M, McKenna WM, Paton MA: A clinical study of Noonan syndrome. Arch Dis Child 1992;67:178–183.

12 Witt DR, Keena BA, Hall JG, Allanson JE: Growth curves for height in Noonan syndrome. Clin Genet 1986;30:150–153.

13 Ranke MB, Heidemann P, Knupfer C, Enders H, Schmaltz AA, Bierich JR: Noonan syndrome: Growth and clinical manifestation in 144 cases. Eur J Pediatr 1988;148:220–227.

14 Tanner JM, Lejarraga H, Cameron N: The natural history of the Silver-Russell syndrome: A longitudinal study of thirty-nine cases. Pediatr Res 1975;9:611–623.

15 Eeckels R, Vanderschueren-Lodeweyckx M, Wolter R: Plasma growth hormone determination in the Silver-Russell syndrome. Helv Paediatr Acta 1970;25:363–370.

16 O'Brien JE, Sadeghi-Nejad A, Feingold M: Growth hormone deficiency in a patient with Silver-Russell syndrome. J Pediatr 1978;93: 152–153.

17 Hall J: Microphallus, growth hormone deficiency and hypoglycemia in Russell-Silver syndrome. Am J Dis Child 1978;132:1149.

18 Stogman W, Borkenstein M, Grubbauer HM: Silver-Russell-Syndrom mit Panhypopituitarismus. Klin Pädiatr 1978;190:586–591.

19 Nishi Y, Kawaguchi S, Nakanishi Y, Usui T: Silver-Russell syndrome and growth hormone deficiency. Acta Paediatr Scand 1982;71: 1035–1036.

20 Theintz G, Alfonso Lopez L, Schorderet D, Sizonenko PC: Growth in a case of Russell-Silver syndrome treated for hypopituitarism. Helv Paediatr Acta 1988;43:325–331.

21 Cassidy SB, Blonder O, Courtney V, Ratzan SK, Carey DE: Russell-Silver syndrome and hypopituitarism. Am J Dis Child 1986;140: 155–159.

22 Draznin MB, Stelling MW, Johansons AJ: Silver-Russell syndrome and craniopharyngioma. J Pediatr 1980;96:887–889.

23 Stanhope R, Ackland F, Hamill G, Clayton J, Jones J, Preece MA: Physiological growth hormone secretion and response to growth hormone treatment in children with short stature and intrauterine growth retardation. Acta Paediatr Scand 1989;349(suppl):47–52.

24 Howard NJ, Moore B, Verge CF, Cowell CT: Growth hormone therapy in Russell-Silver syndrome. Acta Paediatr Scand 1991;379(suppl): 171.

25 Chatelain PG: Auxology and response to growth hormone treatment of patients with intrauterine growth retardaton or Silver-Russell syndrome: Analysis of data from the Kabi Pharmacia International Growth Study. Acta Paediatr 1993;391(suppl):79–81.

26 Lenko HL, Leisti S, Perheentupa J: The efficacy of growth hormone in different types of growth failure. Eur J Pediatr 1982;138:241–249.

27 Butler MG, Meaney FJ: Standards for selected anthropometric measurements in Prader-Willi syndrome. Pediatrics 1991;88:853–859.

28 Lee PDK, Wilson DM, Rountree L, Hintz RL, Rosenfeld RG: Linear growth response to exogenous growth hormone in Prader-Willi syndrome. Am J Med Genet 1987;28:865–871.

29 Tolis G, Lewis W, Verdy M, Friesen HG, Solomon S, Pagalis G, Pavlatos F, Fessas PH, Rochefort G: Anterior pituitary function in the Prader-Labhart-Willi syndrome. J Clin Endocrinol Metab 1974;39:1061–1066.

30 Bray GA, Dahms WT, Swerdloff RS, Fiser RH, Atkinson RL, Carrel RE: The Prader-Willi syndrome: A study of 40 patients and a review of the literature. Medicine 1983;62:59–78.

31 Cacciari E, Cicognani A, Pirazzoli P, Tassoni P, Zappulla F, Salardi S, Bernardi F: Relationships among the secretion of ACTH, GH and cortisol during the insulin-induced hypoglycemia test in normal and obese child. J Clin Endocrinol Metab 1975;40:802–806.

32 Costeff H, Holm VA, Ruvalcaba R, Shaver J: Growth hormone secretion in Prader-Willi syndrome. Acta Paediatr Scand 1990;79:1059–1062.

33 Cappa M, Grossi A, Borrelli P, Ghigo E, Bellone J, Benedetti S, Carta D, Loche S: Growth hormone (GH) response to combined pyridostigmine and GH-releasing hormone administration in patients with Prader-Labhard-Willi syndrome. Horm Res 1993;39:51–55.

34 Loche S, Pintor C, Cappa M, Puggioni R, Locatelli V, Muller EE: Pyridostigmine counteracts the blunted growth hormone response to growth hormone-releasing hormone of obese children. Acta Endocrinol 1989;120:624–628.

35 Raiti S, Trias E, Levitsky L, Grossman MS: Oxandrolone and human growth hormone. Am J Dis Child 1973;126:597–600.

36 Fesseler WH, Bierich JR: Untersuchungen beim Prader-Labhart-Willi-Syndrom. Monatsschr Kinderheilkd 1983;131:844–847.
37 Nilsson KO: What is the value of growth hormone treatment in short children with specified syndrome? Acta Paediatr Scand 1989; 362(suppl):61–68.
38 Mullis PE, Patel MS, Brickell PM, Hindmarsh PC, Brook CGD: Growth characteristics and response to growth hormone therapy in patients with hypochondroplasia: Genetic linkage of the insulin-like growth factor I gene at chromosome 12q23 to the disease in a subgroup of these patients. Clin Endocrinol 1991;34:265–274.
39 Le Merrer M, Rousseau F, Legeai-Mallet L, Landais JC, Pelet A, Bonaventure J, Sanak M, Wessenbach J, Stoll C, Munnich A, Maroteaux P: A gene for achondroplasia-hypochondroplasia maps to chromosome 4p. Nat Genet 1994; 6:318–321.
40 Velinov M, Slaugenhaupt SA, Stoilov I, Scott CI, Gusella JF, Tsipouras P: The gene for achondroplasia maps to the telomeric region of chromosome 4p. Nat Genet 1994;6:314–317.
41 Kodama H, Okabe I, Yanagisawa M: Therapeutic benefit of growth hormone in achondroplastic dwarfism. Acta Paediatr Jpn 1990;32: 323–334.
42 Okabe T, Nishikawa K, Miyamori C, Sato T: Growth-promoting effect of human growth hormone on patients with achondroplasia. Acta Paediatr JPN 1991;33:357–362.
43 Appan S, Laurent S, Chapman M, Hindmarsh PC, Brook CGD: Growth and growth hormone therapy in hypochondroplasia. Acta Paediatr Scand 1990;79:796–803.
44 Horton WA, Hecht JT, Hood OJ, Marshall RN, Moore WV: Growth hormone therapy in achondroplasia. Am J Med Genet 1992;42: 667–670.
45 Bridges NA, Hindmarsh PC, Brook CGD: Growth of children with hypochondroplasia treated with growth hormone for up to three years. Horm Res 1991;36(suppl 1):56–60.
46 Torrado C, Bastian W, Wisniewski K, Castells S: Treatment of children with Down syndrome and growth retardation with recombinant human growth hormone. J Pediatr 1991;119: 478–483.
47 Anneren G, Gustavson KH, Sara VR, Tuvemo T: Growth retardation in Down syndrome in relation to insulin-like growth factors and growth hormone. Am J Med Genet 1990(suppl 7):59–62.
48 Ragusa L, Romano C, Failla P, Proto C, Colabucci F: Growth hormone subnormality in Down syndrome. Am J Med Genet 1992;43: 894–895.
49 Anneren G, Gustafsson J, Sara VR, Tuvemo T: Normalized growth velocity in children with Down's syndrome during growth hormone therapy. J Intellect Disabil Res 1993;37:381–387.
50 Allen DB, Frazier SD, Foley TP, Pescovitz OH: Growth hormone for children with Down syndrome. J Pediatr 1993;123:742–743.
51 Editorial: Little bigger, little better. Lancet 1994;344:627–628.
52 Togoe T, Ohdo S, Hayakawa K: Growth hormone treatment in congenital anomaly syndrome. Acta Paediatr Scand 1991;372(suppl): 153.
53 Larizza D, Maghnie M: Poland's syndrome associated with hormone deficiency. J Med Genet 1990;27:53–55.
54 Kristjansson K, Hoffman WH, Flannery DB, Cohen MJ: Johanson-Blizzard syndrome and hypopituitarism. J Pediatr 1988;113:851–853.
55 Massa G, Dooms L, Vanderschueren-Lodeweyckx M: Growth hormone deficiency in a girl with the Cohen syndrome. J Med Genet 1991;28:48–50.
56 Pivnick EK, Burstein S, Sid Wilroy RA, Kaufman RA, Ward JC: Hallermann-Streiff syndrome with hypopituitarism contributing to growth failure. Am J Med Genet 1991;41:503–507.
57 Schwartz ID, Schwartz KJ, Kousseff BG, Bercu BB, Root AW: Endocrinopathies in Cornelia de Lange syndrome. J Pediatr 1990;117:920–923.
58 Martinez EL, Leon Sanz M, Hawkins Carranzo F: Growth hormone deficiency associated with hereditary multiple exostoses – Growth hormone treatment of one case. Acta Paediatr Scand 1988;343(suppl):218–219.
59 Kodama M, Fujimoto S, Namikawa T, Matsuda I: Aarskog syndrome with isolated hormone deficiency. Eur J Pediatr 1981;135:273–276.
60 Vanderschueren-Lodeweyckx M, Fryns JP, Van den Berge H, et al: Bloom's syndrome. Am J Dis Child 1984;138:812–816.
61 Zucchini A, Manfredi G, Azzaroli L, Manfredi R: Biosynthetic growth hormone in the treatment of Leri-Weill's dyschondrosteosis. Acta Paediatr Scand 1990;370(suppl):195.
62 Thuestad IJ, Ivarsson SA, Nilsson KO: Leri-Weill's syndrome with growth hormone deficiency – Effect of growth hormone treatment. Acta Paediatr Scand 1992;383(suppl):108.
63 Lenko HL, Leisti S, Perheentupa J: The efficacy of growth hormone in different types of growth failure. Eur J Pediatr 1982;138:241–249.
64 Abusrewil SS, McDermott A, Savage DCL: Growth hormone, suspected gonadotropin deficiency, and ring 18 chromosome. Arch Dis Child 1988;63:1090–1091.
65 Meloni A, Boccone L, Angius L, Loche S, Falchi AM, Cao A: Hypothalamic growth hormone deficiency in a patients with ring chromosome 18. Eur J Pediatr 1994;153:110–112.
66 Malpuech CH, Palcoux MC, Boucher D, Cassagne J: Syndrome LEOPARD et insuffisance hypophysaire. Arch Fr Pédiatr 1979;36:413–417.
67 Pochedly C, Collipp PJ, Wolman SR, Suwansirikul S, Rezvani I: Fanconi's anemia with growth hormone deficiency. J Pediatr 1971;79: 93–96.
68 Zachmann M, Illig R, Prader A: Fanconi's anemia with isolated growth hormone deficiency. J Pediatr 1972;80:159–160.
69 Gleadhill V, Bridges JM, Hadden DR: Fanconi's aplastic anaemia with short stature. Arch Dis Child 1975;50:318–320.
70 Boscherini B, Galasso C, Manca Bitti ML: Abnormal face, congenital absence of the left pericardium, mental retardation, and growth hormone deficiency. Am J Med Genet 1994;49: 111–113.
71 Chitayat D, Hall JG, Couch RM, Phang MS, Baldwin VJ: Syndrome of mental retardation, facial anomalies, hypopituitarism, and distal arthrogryposis in sibs. Am J Med Genet 1990; 37:65–70.
72 Stratton RF, Parker MW: Growth hormone deficiency, wormian bones, dextrocardia, brachycamtodactyly, and other midline defects. Am J Med Genet 1989;32:169–173.
73 Haraldsson A, van der Burgt CJAM, Weemaes CMR, Otten B, Bakkeren JAJM, Stoelinga GBA: Antibody deficiency and isolated growth hormone deficiency in a girl with Mulibrey nanism. Eur J Pediatr 1993;152:509–512.

Horm Res 1995;44(suppl 3):49–54

Use of Combined Gn-RH Agonist and hGH Therapy for Better Attaining the Goals in Precocious Puberty Treatment

Luciano Tatò[a]
Giuseppe Saggese[b]
Luciano Cavallo[c]
Franco Antoniazzi[a]
Andrea Corrias[d]
Anna Maria Pasquino[e]
Mariangela Cisternino[f]

[a] Pediatric Clinic, University of Verona;
[b] Pediatric Clinic, University of Pisa;
[c] Pediatric Clinic, University of Bari;
[d] Endocrine Pediatric Division, Torino;
[e] Pediatric Clinic, University 'La Sapienza', Roma;
[f] Pediatric Clinic, IRCCS Policlinico S. Matteo, University of Pavia, Italy

Abstract

We studied 30 girls (age 6.36 ± 1.21 years, range 4.6–8.8) affected by idiopathic precocious puberty with significant reduction of height velocity (below the 25th centile) at the end of 1 year of Gn-RHa (triptorelin intramuscular depot) treatment, to evaluate GH-IGF-I axis activity and the effects of combined Gn-RHa plus hGH therapy. After 12 months, 15 patients continued Gn-RHa and started hGH therapy for 12 months, while 15 continued treatment with Gn-RHa alone (control group). We evaluated height velocity, bone age, urinary GH, serum IGF-I and IGFBP-3 levels throughout the study; plasma GHBP levels were determined only in the first 12 months of Gn-RHa treatment. Height velocity decreased significantly during Gn-RHa treatment; it increased significantly and became higher than the control group after 12 months of Gn-RHa plus hGH treatment. During Gn-RHa therapy alone, bone age progressed less than chronological age, while in the 12 months of Gn-RHa plus hGH treatment there was a slight nonsignificant increase in bone age progression in comparison to controls. Serum IGF-I and IGFBP-3 levels decreased significantly at 12 months of Gn-RHa therapy and increased significantly after Gn-RHa plus hGH treatment. Urinary GH levels showed the same behavior. Plasma GH binding to peak II-BP, slightly lower than the prepubertal normal range before treatment, significantly increased after 12 months of Gn-RHa treatment. Therefore, in these girls, during Gn-RHa treatment alone, we have a reduction in GH-IGF-I axis activity. During Gn-RHa plus hGH therapy there was a significant increase in height velocity, in urinary GH levels, in serum IGF-I and IGFBP–3 levels. Bone age did not seem to advance faster than chronological age and this may imply a better prediction in adult height.

In our opinion, only in a small percentage of patients affected by precocious puberty (with a very low predicted adult height or an important reduction of growth velocity during Gn-RHa treatment) may an association with hGH therapy be useful.

Key Words
Precocious puberty
Gonadotropin-releasing hormone agonists
hGH therapy

Prof. Luciano Tatò
Clinica Pediatrica, Università di Verona
Policlinico Borgo Roma
I–37134 Verona (Italy)

0301–0163/95/0449–0049
$8.00/0

Introduction

The main aims of therapy in precocious puberty are to inhibit puberty, to stop and possibly reverse the progression of secondary sex characteristics and particularly menses, to slow down skeletal maturation to delay epiphyseal fusion and consequently to improve final height.

In untreated patients with precocious puberty final adult height is severely compromised if compared to target height [1, 2]; progestational drugs, effective in arresting the advancement of pubertal signs, have been abandoned because of their weak action on statural and bone age progression, with a poor result on final height [3, 4].

Gonadotropin-releasing hormone agonists (Gn-RHa) are now the therapy of choice when the disease is gonadotropin-releasing hormone (Gn-RH) dependent [5]. They are effective in arresting pubertal progression: secondary sex characteristics regress during the first year, even if they do not disappear completely [6]. After the end of therapy, there is a rapid progression of puberty with a development of regular ovulatory menstrual function [7].

During treatment, a decrease in both growth velocity rate and bone age maturation is observed, with an improvement in predicted adult height at the end of treatment compared to the start of therapy. In effect, treatment delays epiphyseal closure and improves final height as reported by many studies [8–12]. The result of reducing growth velocity and arresting pubertal development is not attained immediately after the beginning of therapy. Growth continues at high levels in the first 3 months and only at the end of the first year is it reduced to prepubertal levels [6]. In some subjects this reduction is very notable and height velocity reaches the level of children with GH deficiency. In effect, Gn-RHa treatment in precocious puberty seems to reduce GH secretion with an amplitude-modulated phenomenon, without significant change in the number of secretory peaks and IGF-I levels [13].

The aims of this study were to evaluate the GH-IGF-I axis activity and the effects of combined Gn-RHa plus hGH therapy in a group of girls with significant reduction of height velocity at the end of 1 year of Gn-RHa treatment alone.

Patients and Methods

Thirty girls (age 6.36 ± 1.21 years, range 4.6–8.8) affected by idiopathic precocious puberty were studied as a part of a multicenter study by the Italian Society of Pediatric Endocrinology and Diabetology including more patients. The diagnosis was based on the appearance of pubertal signs (B < 2 according to Tanner) before 8 years of age, bone age (BA) > 1 year of chronological age (CA), LH and FSH response to the Gn-RH stimulation test (100 μg/m^2 i.v. bolus dose) in the pubertal range [14], uterus longitudinal diameter (detected by ultrasonography) > 3.5 cm.

No patients had any evidence of progressive organic central nervous system disorders at computed tomography (CT) or magnetic resonance imaging (MRI), thyroid or GH deficiency (GH response to clonidine stimulation test: 150 μg/m^2 or > 10 ng/ml), identifiable adrenal or gonadal pathology, hepatic or renal problems, or had previous treatment with inhibitory steroids.

Girls received Gn-RHa (triptorelin) in a depository intramuscular formulation (Decapeptyl–Ipsen) at the dose of 60 μg/kg every 28 days (only the interval between first and second administration was of 21 days) for a 24-month period. The initial doses were adjusted during the study according to weight. Informed consent was obtained from the parents of each girl before starting therapy.

The suppression of pituitary gonadotropins was controlled every 6 months by measurements of serum estradiol level and Gn-RH stimulation test (100 μg/m^2 i.v. bolus dose). These measurements were taken 26 days after the intramuscular Gn-RHa administration.

Desensitization of pituitary was considered adequate when there was no significant increase of LH and FSH to Gn-RH stimulation test (LH and FSH peak < 2 IU/l) and basal serum estradiol levels were < 15 pg/ml.

From the 9th to 12th month of Gn-RHa treatment the 30 patients had a reduction of height velocity rate below the 25th centile calculated for the patient's CA and BA [15] and were admitted to the second part of the study protocol.

After 12 months of treatment with Gn-RHa, 15 patients continued treatment with Gn-RHa and started hGH therapy (Saizen, Serono) at the dosage of 0.7 IU/kg/week, administered in 6 injections/week subcutaneously in the evening for 12 months, while 15 continued to be treated with Gn-RHa alone and were considered as a control group. Patients were randomly assigned to treatment with Gn-RHa plus hGH or treatment with Gn-RHa alone.

Standing height was taken as the average of three measurements obtained in the morning by a wall-mounted Harpenden stadiometer. Height velocity was calculated in the 3 months before Gn-RHa treatment, and every 3 months after. BA was determined every 6 months during the study according to the method of Greulich and Pyle [16].

LH and FSH levels were determined using fluorimetric enzyme immunoassay (Baxter Diagnostics, Deerfield, Ill., USA); estradiol levels were measured using radio-immunoassay (Sorin Biomedica, Saluggia, Italy); GH levels were determined by immunoradiometric assay (Biodata, Guidonia Montecelio, Italy).

Serum IGF-I and urinary GH excretion were evaluated every 3 months during the study; serum IGFBP–3 levels were assayed every 6 months; plasma GHBP levels were determined every 6 months only during the first 12 months of Gn-RHa treatment. All serum IGF-I, IGFBP-3 and plasma GHBP samples and urine for GH determination were stored at –80 °C until the assay. All assays, except GHBP, were performed with commercial kits.

Serum IGF-I levels were measured by immunoradiometric assay after acid extraction (Diagnostic Systems Laboratories, Webster, Tex., USA). The sensitivity of the assay was 7.0 ng/ml. The intra- and interassay coefficients of variation were < 7.2% and < 16.3% respectively. Our normal range in a group of 15 prepubertal girls of the same age, without short stature, was 385.3 ± 44.1 ng/ml.

Serum IGFBP-3 levels were measured by radio-immunoassay (Diagnostic Systems Laboratories). The sensitivity of the assay was

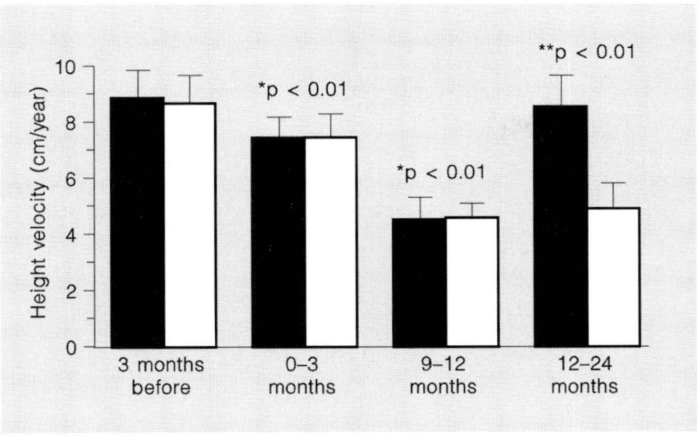

Fig. 1. Height velocity (cm/year) in girls with precocious puberty treated with Gn-RHa intramuscular depot for 12 months and subsequently with Gn-RHa plus hGH for 12 months (■). Control group represents girls treated only with Gn-RHa (□) (mean ± SD). * $p < 0.01$ vs. before; ** $p < 0.01$ vs. 9–12 months and vs. control group.

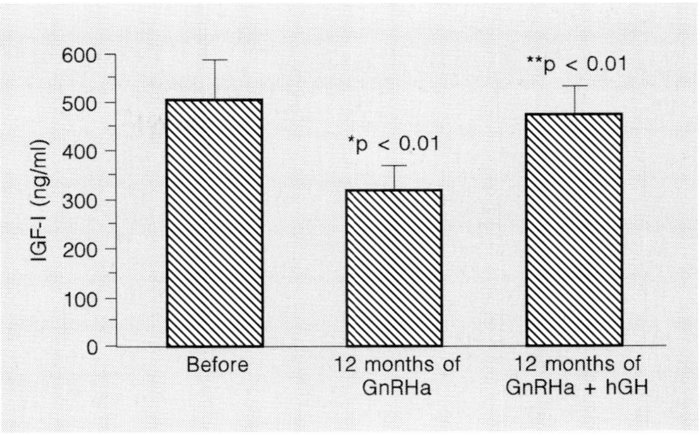

Fig. 2. IGF-I serum levels (ng/ml) in girls with precocious puberty before the start of the study, at 12 months of Gn-RHa treatment and after 12 months of Gn-RHa + hGH treatment (mean ± SD). * $p < 0.01$ vs. before; ** $p < 0.01$ vs. 12 months of Gn-RHa.

0.9 ng/ml. The intra- and interassay coefficients of variation were <6.7% and <8.2% respectively. The normal range in the same group of prepubertal girls was 3.38 ± 0.6 mg/l.

GH urinary levels were measured in overnight urinary samples by immunoradiometric assay (Bio-Mérieux, Marcy l'Etoile, France). The sensitivity of the assay was 0.5 pg/ml. The intra- and interassay coefficients of variation were less than <6.6% and <8.1% respectively. Results were expressed in ng/g of creatinine in the same urinary sample assayed with routine methods. The normal range in the same group of prepubertal girls was 22.82 ± 3.3 ng/g creatinine. To eliminate the interassay variation, all samples were measured in duplicate in the same assay.

GHBP was measured in plasma by HPLC gel filtration by the method of Tar et al. [17]. The GH binding was expressed as the percentage of total radioactivity (^{125}I-hGH) bound to peak I-BP (low affinity complex), to peak II-BP (main GH-BP) and III (free ^{125}I-hGH). The normal GH binding in the group of prepubertal girls of the same age were 2.5 ± 0.3% for peak I-BP and 23.6 ± 2.5% for peak II-BP.

Statistical analyses were performed using unpaired and paired two-tailed Student's t test and ANOVA for repeated measures. All data were expressed as mean ± SD.

Results

Height velocity was 8.87 ± 1.01 cm/year in the 3 months before starting treatment; during Gn-RHa treatment it decreased significantly to 4.47 ± 0.76 cm/year from the 9th to 12th month ($p < 0.01$); after 12 months of Gn-RHa plus hGH treatment, height velocity increased to 8.58 ± 1.12 cm/year ($p < 0.01$ vs. 9–12 months Gn-RHa period and vs. controls) (fig. 1).

During the 12 months of Gn-RHa therapy alone, BA progressed less than CA (ΔBA/ΔCA 0.82 ± 0.11 year; control group 0.85 ± 0.10 year). In the 12 months of Gn-RHa plus hGH treatment there was a slight nonsignificant increase in BA progression in comparison to controls (ΔBA/ΔCA 0.91 ± 0.12 year; control group 0.83 ± 0.15 year; NS).

Serum IGF-I levels were higher, but not significantly, than the prepubertal normal range and decreased significantly from 503.1 ± 84.2 to 318.3 ± 51.5 ng/ml at 12 months of Gn-RHa treatment ($p < 0.01$); after 12 months of Gn-RHa plus hGH treatment, IGF-I levels increased significantly to 470.4 ± 72.0 ng/ml ($p < 0.01$) (fig. 2).

Serum IGFBP-3 levels were higher than the prepubertal normal range before Gn-RHa and decreased significantly from 5.23 ± 1.15 to 2.98 ± 0.64 mg/l at 12 months of Gn-RHa treatment ($p < 0.01$); after 12 months of Gn-RHa plus hGH therapy, IGFBP-3 increased significantly to 5.91 ± 1.41 mg/l ($p < 0.01$) (fig. 3).

Urinary GH levels were higher than the normal range before treatment, but not significantly (27.51 ± 8.13 vs. 22.82 ± 5.3 ng/g creatinine; NS). During Gn-RHa treatment, GH urinary levels fell dramatically already at 6 months and had a further decrease at 12 months (10.75 ± 1.42 and 6.88 ± 1.3 ng/g creatinine respectively; $p < 0.01$ in both cases). During Gn-RHa plus hGH therapy there was the expected increase of urinary GH levels, that testified the hGH administration (fig. 4).

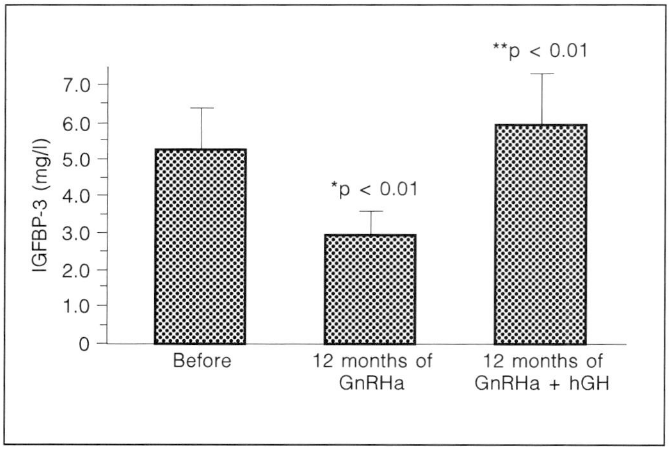

3

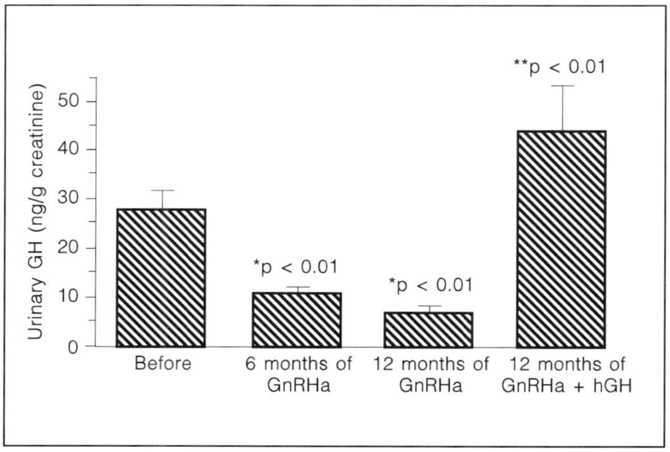

4

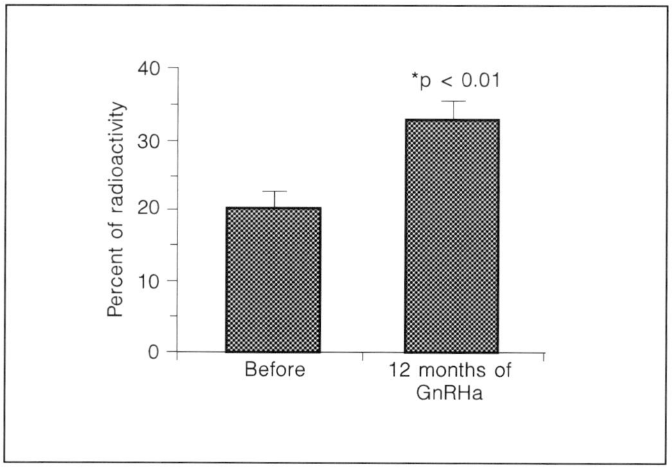

5

Fig. 3. IGFBP-3 serum levels (mg/l) in girls with precocious puberty before the start of the study, at 12 months of Gn-RHa treatment and after 12 months of Gn-RHa + hGH treatment (mean ± SD). * $p < 0.01$ vs. before; ** $p < 0.01$ vs. 12 months of Gn-RHa.

Fig. 4. GH levels in overnight urine (ng/g creatinine) in girls with precocious puberty before the start of the study, at 6 and 12 months of Gn-RHa treatment and after 12 months of Gn-RHa + hGH treatment (mean ± SD). * $p < 0.01$ vs. before; ** $p < 0.01$ vs. before, 6 and 12 months of Gn-RHa.

Fig. 5. GH binding to II-BP (^{125}I-hGH bound to peak II-BP: % of total radioactivity) in girls with precocious puberty before the start of the study and at 12 months of Gn-RHa treatment (mean ± SD). * $p < 0.01$ vs. before.

Plasma GH binding activity in the patients was slightly lower than the prepubertal normal range before treatment (binding of labelled hGH to peak II-BP 20.13 ± 1.48%, NS; binding to I-BP 2.2 ± 0.2%, NS); GH binding to peak II-BP significantly increased after 12 months of Gn-RHa treatment (II-BP 32.30 ± 2.68, $p < 0.01$; I-BP 1.9 ± 0.3%, NS) (fig. 5).

Discussion

During Gn-RHa treatment, a decrease in both growth velocity rate and BA maturation were observed, with an improvement in predicted adult height at the end of treatment compared to the start of therapy. In effect, treatment delays epiphyseal closure and improves final height as reported by many studies [8–12]. However, height potential also seems to be preserved in girls with slowly progressive untreated precocious puberty [18]. There is, therefore, a beneficial effect of treatment on final height, particularly when therapy is begun soon after the onset of precocious puberty and there is only a slight advancement of bone age [10–12].

In a recent Italian multicenter study, two groups of girls affected by idiopathic precocious puberty, treated with two different Gn-RHa, triptorelin used intramuscularly every 28 days and buserelin used as intranasal spray, were compared [19]. Treated patients showed an improvement in final height if compared to untreated patients and to predicted height at the start of treatment, with both agonist treatments. Final height was lower than predicted height at the end of treatment, as reported by others [11]. However, the group treated with depository intramuscular agonist preparation had a better improvement in adult height and reached or exceeded target height.

In this study, in the Gn-RHa-treated patients there was an important reduction in growth velocity rate, below the

25th centile calculated for patient's CA and BA. In these girls, during Gn-RHa treatment alone, we observed a reduction in GH-IGF-I axis activity, testified by the decrease of urinary GH, IGF-I and IGFBP-3 serum levels, and by the increase of plasma GHBP.

IGF-I decreased in a significant way at 12 months of Gn-RHa treatment alone. This is not in agreement with other studies conducted for a shorter period and with a lower decrease in growth velocity [20]. However our patients were a selected group of girls in which significantly decreased growth velocity was an admission criteria to the study protocol.

IGFBP-3 showed the same pattern as IGF-I, while urinary GH decreased significantly after 6 months of treatment and had a further decrease after 12 months. This is not strange, seeing that all these parameters are correlated positively with GH-IGF-I axis activity [21, 22]. On the other hand, GHBP levels, that are regulated positively from GH and negatively from sex hormone levels, were lower than controls before therapy and showed a rise during Gn-RHa treatment, as reported by others [23].

During Gn-RHa plus hGH therapy there was a significant increase in height velocity according to other authors that treated patients with combined true precocious puberty and GH deficiency [24] or precocious puberty alone [25]. Moreover, we showed an increase in serum IGF-I and IGFBP-3 levels, and in urinary GH levels, as expected. BA did not seem to advance faster than CA and this may imply a better prediction in adult height.

In conclusion, Gn-RH agonists have been demonstrated to be effective in improving final adult height in patients affected by precocious puberty, especially if given as depository formulations. In selected cases with an important reduction of growth velocity or with a predicted adult height falling significantly, an association with hGH therapy may be useful. However, in our opinion, the good results in final height obtained with Gn-RHa therapy alone limits this treatment only to a small percentage of cases.

References

1 Thamdrup E: Precocious Sexual Development: A Clinical Study of 100 Children. Springfield, Thomas, 1961, pp 44–63.

2 Sigurjonsdottir TH, Hayles AB: Precocious puberty. A report of 96 cases. Am J Dis Child 1968;115:309–321.

3 Cisternino M, De Luca F, Tatò L, Antoniazzi F, Balducci R, Bartolotta E, Bernasconi S, Chiumello G, Corrias A, De Sanctis C, De Sanctis V, Marchione SA, Meroni R, Nizzoli G, Pasquino AM, Pucarelli I, Pozzan GB, Ventura M, Severi F: Statura definitiva di pazienti affetti da pubertà precoce vera trattate con ciproterone acetato: Studio multicentrico italiano. Riv Ital Pediatr 1993;19:48–53.

4 Cisternino M, Pasquino AM, Bozzola M, Balducci R, Lorini R, Pucarelli I, Segni M, Severi F: Final height attainment and gonadal function in girls with precocious puberty treated with cyproterone acetate. Horm Res 1992;37:86–90.

5 Kaplan SL, Grumbach MM: Clinical review: Pathophysiology and treatment of sexual precocity. J Clin Endocrinol Metab 1990;71:785–789.

6 Tatò L, Cisternino M, Bozzola M, Romano MR, Antoniazzi F, Biscaldi I, Spolettini E, Roger M, Chaussain JL, Gaburro D, Severi F: Il trattamento della pubertà precoce idiopatica con un analogo del LH-RH (*D*-Trp-LHRH) ad azione prolungata. Riv Ital Pediatr 1988;14:52–57.

7 Jay N, Mansfield MJ, Blizzard RM, Crowley WF Jr, Schoenfeld D, Rhubin L, Boepple PA: Ovulation and menstrual function of adolescent girls with central precocious puberty after therapy with gonadotropin-releasing hormone agonists. J Clin Endocrinol Metab 1992;75:890–894.

8 Kauli R, Kornreich L, Laron Z: Pubertal development, growth and final height in girls with sexual precocity after therapy with the GnRH analogue *D*-Trp-6-LHRH. A report on 15 girls, followed after cessation gonadotrophin suppressive therapy. Horm Res 1990;33:11–17.

9 Oostdijk W, Drop SL, Odink RJ, Hummelink R, Partsch CJ, Sippell WG: Long-term results with a slow-release gonadotrophin-releasing hormone agonist in central precocious puberty. Dutch-German Precocious Puberty Study Group. Acta Paediatr Scand 1991(suppl 372):39–45.

10 Boepple PA: Long-term experience with GnRH agonist treatment of central precocious puberty. Clin Pediatr Endocrinol 1993;2(suppl 3):49–56.

11 Oerter KE, Manasco PK, Barnes KM, Jones J, Hill S, Cutler GB Jr: Effects of luteinizing hormone-releasing hormone agonists on final height in luteinizing hormone-releasing hormone-dependent precocious puberty. Acta Paediatr 1993(suppl 388):62–68.

12 Chaussain JL, Bost M, Roger M: Taille définitive des pubertés précoces. Arch Pédiatr 1994;1:368–370.

13 Di Martino-Nardi J, Wu R, Fishman K, Saenger P: The effect of long-acting analog of luteinizing hormone releasing hormone on growth hormone secretory dynamics in children with precocious puberty. J Clin Endocrinol Metab 1991;73:902–906.

14 Oerter KE, Uriarte MM, Rose SR, Barnes KM, Cutler GB Jr: Gonadotropin secretory dynamics during puberty in normal girls and boys. J Clin Endocrinol Metab 1990;71:1251–1258.

15 Tanner JM, Whitehouse RH: Clinical longitudinal standards for height, weight, height velocity, weight velocity and stages of puberty. Arch Dis Child 1976;51:170–179.

16 Greulich WW, Pyle SI: Radiologic Atlas of Skeletal Development of the Hand and Wrist, ed 2. Stanford, Stanford University Press, 1959.

17 Tar A, Hocquette JF, Souberbielle JC, Clot JP, Brauner R, Postel-Vinay MC: Evaluation of the growth hormone-binding proteins in human plasma using high pressure liquid chromatography gel filtration. J Clin Endocrinol Metab 1990;71:1202–1207.

18 Kreiter M, Burstein S, Rosenfield RL, Moll GW Jr, Cara JF, Yousefzadeh DK, Cuttler L, Levitsky LL: Preserving adult height potential in girls with idiopathic true precocious puberty. J Pediatr 1990;117:364–70.
19 Antoniazzi F, Cisternino M, Nizzoli G, Bozzola M, Corrias A, De Luca F, De Sanctis C, Rigon F, Zamboni G, Bernasconi S, Chiumello G, Severi F, Tatò L: Final height in girls with central precocious puberty. Comparison of two different luteinizing hormone-releasing hormone agonist treatments. Acta Paediatr 1994; 83.
20 Sklar CA, Rothenberg S, Blumberg D, Oberfield SE, Levine LS, David R: Suppression of the pitutary-gonadal axis in children with central precocious puberty: Effects on growth, growth hormone, insulin-like growth factor-I, and prolactin secretion. J Clin Endocrinol Metab 1991;73:734–738.
21 Blum WF, Ranke MB: Insulin-like growth factor binding proteins (IGFBPs) with special reference to IGFBP-3. Acta Paediatr Scand 1990; 367(suppl):55–62.
22 Albini CH, Quattrin T, Vandlen RL, MacGillivray MH: Quantitation of urinary growth hormone in children with normal and abnormal growth. Pediatr Res 1988;23:89–92.
23 Postel-Vinay MC, Fontoura M: Regulation of growth hormone binding proteins in human plasma. Acta Paediatr Scand 1991;379(suppl): 79–82.
24 Cara JF, Kreiter ML, Rosenfield RL: Height prognosis of children with true precocious puberty and growth hormone deficiency: Effect of combination therapy with gonadotropin releasing hormone agonist and growth hormone. J Pediatr 1992;120:709–715.
25 Darendeliler F, Hindmarsh PC, Brook CGD: Non-conventional use of growth hormone: European experience. Horm Res 1990;33:128–136.

Horm Res 1995;44(suppl 3):55–63

Giuseppe Saggese
Giampiero I. Baroncelli
Giovanni Federico
Silvano Bertelloni

Endocrine Unit, Chair of Preventive Pediatrics, Department of Pediatrics, University of Pisa, Italy

Effects of Growth Hormone on Phosphocalcium Homeostasis and Bone Metabolism

Key Words
Bone density
Bone turnover
Chondrocytes
Collagen metabolism
Growth hormone
Osteoblast-like cells
Phosphocalcium metabolism
Vitamin D

Abstract
In this review the effects of growth hormone (GH) on phosphocalcium homeostasis and bone metabolism are reported. Some in vitro effects of GH on chondrocytes and osteoblasts are discussed too. The main GH effects on phosphocalcium homeostasis are the permissive action on renal 1α-hydroxylase activity by the hypophosphatemic stimulus and the antiphosphaturic effect by the stimulation of the maximum rate of renal tubular reabsorption of phosphate. On bone, GH is able to stimulate bone turnover and to increase bone mass. In addition, GH stimulates type I and type III collagen metabolism. In vitro, GH increases the proliferation of chondrocytes from the human growing cartilage together with the levels of interleukin-6 in the supernatant. The hormone increases also the proliferation of the human osteosarcoma-derived osteoblast-like cells and augments the osteocalcin levels in the supernatant. Thus, GH markedly influences phosphocalcium homeostasis and bone metabolism in childhood and adolescence. In addition, it is possible that GH continues to play a role in bone physiology during adulthood when final height is reached.

Growth hormone (GH) acts on epiphyseal growth plate either directly or indirectly through the insulin-like growth factor-I (IGF-I) [1]. Furthermore, GH has enteral and renal effects that markedly influence phosphocalcium homeostasis and parathyroid hormone (PTH)-vitamin D endocrine system [2, 3].

In order to evaluate bone metabolism and the growth-related changes, serum measurement of some biochemical markers has been proposed. Of these, osteocalcin and type I and type III procollagen propeptides seem to be promising to study the effects of GH in normal children and in patients with short stature related or not to GH deficiency [4]. Moreover, some evidence suggests that GH has a role in the normal physiology of adult bone after the cessation of longitudinal growth [3].

Effects of GH on Phosphocalcium Metabolism

The effects of GH on phosphocalcium metabolism have been ascribed to a direct effect on intestine and kidney and to the action on PTH-vitamin D endocrine system [2, 5].

Prof. Giuseppe Saggese, MD
Chair of Preventive Pediatrics
Department of Pediatrics, University of Pisa
Via Roma, 35
I-56125 Pisa (Italy)

0301–0163/95/0449–0055
$8.00/0

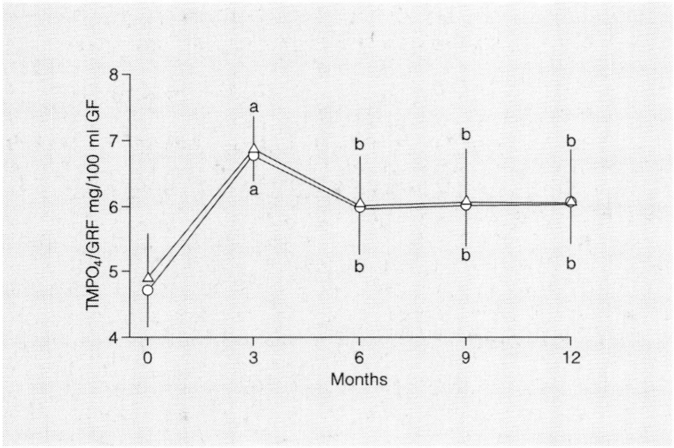

Fig. 1. Variations of $TMPO_4/GFR$ ratio (mean ± SD) in children with classic GH deficiency (○) and nonclassic GH deficiency (△) during GH treatment. [a] $p < 0.01$; [b] $p < 0.03$ vs. 0. [Reproduced from 9, with permission.]

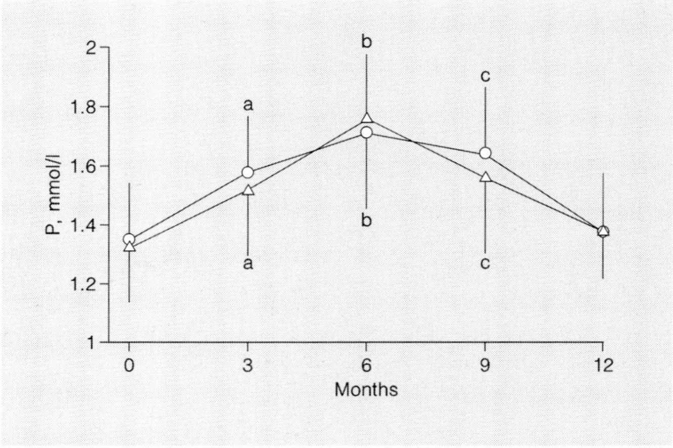

Fig. 2. Variations of serum phosphate concentrations (mean ± SD) in children with classic GH deficiency (○) and nonclassic GH deficiency (△) during GH treatment. [a] $p < 0.03$; [b] $p < 0.001$; [c] $p < 0.02$ vs. 0. [Reproduced from 9, with permission.]

Before GH treatment, intestinal tract is relatively insensitive to 1,25-dihydroxyvitamin D ($1,25(OH)_2D$) action with respect to calcium transport [6]. During GH administration, balance studies showed that a net calcium retention occurs, with stimulation of intestinal calcium absorption and decreased fecal excretion [2, 7]. Chipman et al. [6] suggested that GH directly stimulates intestinal calcium absorption restoring the normal integrity of intestinal mucosa and inducing heightened sensitivity to $1,25(OH)_2D$. Furthermore, in the small intestine of hypophysectomized rats, reduced levels of vitamin D-dependent calcium binding protein have been reported, as well as their increase after GH treatment [8]. The effects of GH on urinary calcium excretion are variable. Indeed, urinary calcium excretion has been reported increased [2, 7], unchanged [6, 9] or decreased [10].

As regards phosphate metabolism, it has been demonstrated that GH supports phosphate retention by increasing the renal threshold for phosphate excretion. Such an effect derives from an increase of maximum tubular phosphate reabsorption rate ($TMPO_4/GFR$) [2, 9]. Indeed, in children with GH deficiency, the values of $TMPO_4/GFR$ have been found in the lower limit of normal [9] or reduced [11], while GH treatment induced a significant increase of $TMPO_4/GFR$ [9, 11] (fig. 1). The effect of GH on $TMPO_4/GFR$ occurs independently of PTH, vitamin D or urinary cyclic AMP, and it is mediated by IGF-I [2, 10, 12–14]. In this regard, the higher $TMPO_4/GFR$ and phosphate values observed during childhood and adolescence in comparison with those in adulthood may reflect the antiphosphaturic effect of GH [15, 16]. Moreover, it may be that during the adolescent growth spurt, a relative phosphate deficiency associated with the increased GH secretion leads to higher $TMPO_4/GFR$ values [2, 16] and $1,25(OH)_2D$ concentrations [17]. Phosphate retention is also stimulated by an augmented intestinal absorption [2, 3]. The intestinal effect of GH on phosphate absorption may be mediated by an increased $1,25(OH)_2D$ production [18] or related to the sensitization of the gut epithelium to vitamin D in the presence of normal amounts of GH [3].

In children with GH deficiency, a significant increase of serum phosphate levels was found at 3 and 6 months of GH treatment; at 9 months phosphate levels decreased even if they remained significantly higher than before. At 12 months of GH treatment, phosphate values were comparable to those we found before treatment [9] (fig. 2). We demonstrated that growth velocity was positively related to $TMPO_4/GFR$ ratio ($r = 0.49$, $p < 0.01$) [9], as also found by Nishiyama et al. [11] ($r = 0.78$, $p < 0.01$). These authors [11] suggested also that $TMPO_4/GFR$ ratio can be used to predict growth velocity after 1 year of GH therapy.

Therefore, the increased renal threshold for phosphate excretion and the increased intestinal calcium and phosphate absorption demonstrate the positive effect of GH on phosphocalcium metabolism.

Effects of GH on Vitamin D Metabolism, PTH, and Calcitonin

In children with GH deficiency, decreased $1,25(OH)_2D$ concentrations have been reported [9, 19], whereas Chipman et al. [6] showed paradoxically elevated serum $1,25(OH)_2D$ levels. The low $1,25(OH)_2D$ levels have been related to a reduced renal 1α-hydroxylase activity because of the absence of the permissive action of GH-IGF-I axis on this enzyme; indeed, GH-IGF-I axis mediates the hypophosphatemic stimulus on renal 1α-hydroxylase [14, 20, 21]. In hypophysectomized rats, the renal $1,25(OH)_2D$ synthesis in response to phosphate deprivation is impaired, such an effect being reversed only by GH [20] or IGF-I [15, 21] administration. In confirmation of these data, in children with GH deficiency during GH replacement therapy, serum $1,25(OH)_2D$ concentrations increased only when a relative hypophospatemia occurred regardless of the presence of serum PTH levels [9, 14]. Burstein et al. [22] also found increased $1,25(OH)_2D$ concentrations during GH treatment, but phosphate levels were not measured. In addition, a rapid increment of $1,25(OH)_2D$ levels during GH administration was showed by Spanos et al. [23] in rats and by Marcus et al. [24] in elderly people. Such a result occurred despite the potent inhibitory effect on renal 1α-hydroxylase activity of the concomitant increase of serum phosphate concentrations. However, decreased [6] or unchanged [10] $1,25(OH)_2D$ levels have been reported in GH-deficient children during GH administration and most cases of acromegaly showed normal $1,25(OH)_2D$ levels too [18]. Anyway, the effect of GH on $1,25(OH)_2D$ production may be only transitory [18], and few long-term data are available on this issue.

As regards the effect of GH on 25-hydroxyvitamin D, no variation of its circulating levels has been observed in children with GH deficiency before and during GH treatment [7, 9, 10, 14, 22].

Increased serum PTH levels have been found during GH administration in rats [25], in GH-deficient children [9], and in elderly people [24]. Such an effect may indicate a stimulatory action of GH on parathyroid gland secretion, as suggested by the increased nephrogenous cyclic AMP values [9, 22]. In addition, no variation of cyclic AMP [7, 10] and PTH values [7, 10, 12, 22] or even decreased [26] PTH levels have been observed during GH administration. Anyway, the increment of cyclic AMP in the absence of an increase in serum PTH levels may suggest that GH sensitizes the kidney to a cyclic AMP-mediated effect of PTH [3, 22]. The consequences on bone mass of the increased PTH levels are unknown. Indeed, the increased serum levels of osteocalcin and carboxyterminal propeptide of type I procollagen (PICP) found during GH treatment in children with GH deficiency (see below) could reflect, at least in part, an increased bone turnover due to an acceleration in remodeling activation frequency, as suggested by Marcus et al. [24].

As regards calcitonin, conflicting data have been reported during GH administration [2]. We [9] and others [22] did not observe changes of serum calcitonin concentrations during 1 year of GH treatment in GH deficiency. Kruse [27] suggested that the stimulated bone turnover and renal phosphate reabsorption may be, at least in part, the result of suppressed calcitonin secretion induced by GH-IGF-I.

So, the main effect of GH-IGF-I axis on PTH-vitamin D endocrine system seems to be the permissive action on the renal 1α-hydroxylase activity, whereas the effects on serum $1,25(OH)_2D$, PTH, and calcitonin levels are variable.

Effects of GH on Bone Turnover and Collagen Metabolism

During childhood, serum osteocalcin, PICP, and aminoterminal propeptide of type III procollagen (PIIINP) concentrations vary with age, with patterns resembling the growth velocity curves; higher values are observed during the first 2 years of life and at puberty, with an earlier increase occurring in girls than in boys [28–33]. Thus, osteocalcin, PICP, and PIIINP levels may have a clinical value as biochemical indexes of growth-related changes during childhood and adolescence. Serum levels of osteocalcin and PICP reflect bone formation, being osteoblast products [34–36]. Osteocalcin, or bone γ-carboxyglutamic acid (Gla) protein, is a major noncollagenous protein of the bone matrix; a small fraction, less than 1%, is released into circulation, where it is measurable by radioimmunoassay [35]. PICP is released into circulation with a stoichiometric ratio of 1:1 versus the type I collagen molecules incorporated into the extracellular matrix [37]. Type I collagen is the most abundant body collagen; it accounts for more than 90% of the organic bone matrix [34, 37], but the same gene product also occurs in soft connective tissues, where type I collagen is present with a mixture of other collagens [37]. Serum PIIINP levels may give information about the activity of connective tissue during growth because type III collagen accounts for 10–50% of total body collagen [38]. Type III collagen is mainly

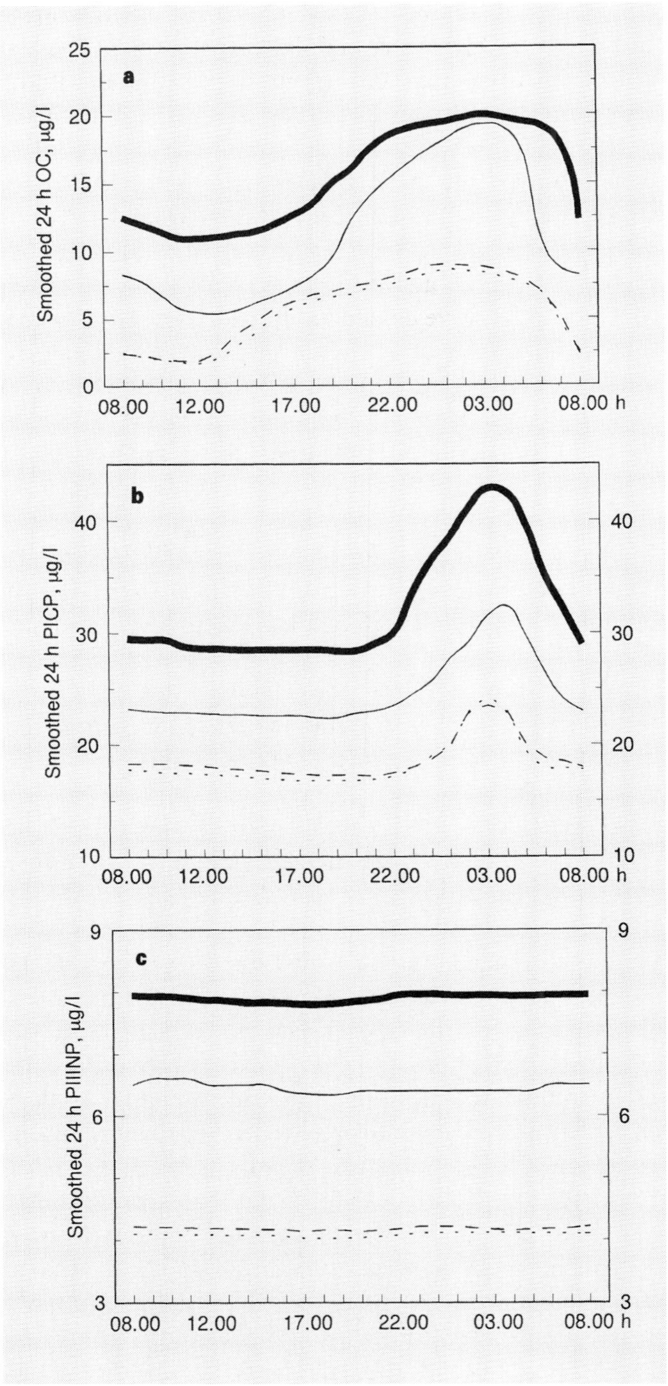

Fig. 3. Pattern of smoothed 24 h osteocalcin (**a**), PICP (**b**) and PIIINP (**c**). Mean data of the subjects. ···· = Classic GH deficiency; --- = nonclassic GH deficiency; —— = short normal children; ▬ = controls. Classic GH deficiency and nonclassic GH deficiency vs. controls: $p < 0.001$ at all the time points of measurement for osteocalcin, PICP, and PIIINP. Short normal children vs. controls: for osteocalcin, $p < 0.01$ from 06:00 to 19:00 h and p = NS from 20:00 to 05:00 H; for PICP, $p < 0.01$ from 08:00 to 23:00 h and $p < 0.001$ from 00:00 to 07:00 h; for PIIINP, $p < 0.001$ at all the time points of measurement. [Reproduced from 4, with permission.]

located in soft connective tissues where it is a major component of the interstitial fibrils [38]. So, in this way, PIIINP levels reflect the somatic growth [4, 37].

Before GH treatment, low serum concentrations of osteocalcin [9, 39–42], PICP [9, 30, 43], bone alkaline phosphatase isoenzyme [44], and PIIINP [30, 45, 46] have been found in children with GH deficiency. Recently, we demonstrated that normal statured prepubertal healthy children, GH-deficient children, and short normal children showed a circadian rhythm of osteocalcin and PICP with higher nocturnal than diurnal values. No variation of serum PIIINP concentrations during daytime and nighttime was observed [4] (fig. 3). The children with GH deficiency had circadian values of osteocalcin, PICP, and PIIINP significantly lower than those of age-matched controls. Short normal children showed a circadian osteocalcin pattern similar to that of controls (fig. 3), whereas the diurnal 12-hour mean values were significantly lower [4]. In these children, circadian values of PICP and PIIINP were significantly lower than controls at all the time points of measurement [4] (fig. 3). Moreover, 24-hour PICP and PIIINP mean data were positively related to spontaneous 24-hour GH concentrations ($r = 0.77$, $p < 0.005$, and $r = 0.69$, $p < 0.005$, respectively) and growth velocity ($r = 0.85$, $p < 0.005$, and $r = 0.70$, $p < 0.005$, respectively) whereas 24-hour osteocalcin mean data did not [4]. In children with GH deficiency, the reduced baseline and circadian osteocalcin concentrations probably reflect a decreased bone turnover and bone formation. In short normal children, the reduced osteocalcin concentrations during daytime suggest that an impaired osteoblasts cyclic activity in producing this bone protein may be present in these patients. The concomitant reduction of baseline and circadian values of PICP and PIIINP in children with GH deficiency and in short normal children revealed that the poor growth was associated with decreased type I and type III collagen synthesis, respectively.

During GH treatment, osteocalcin, PICP, bone alkaline phosphatase isoenzyme, and PIIINP levels progressively increased in GH deficiency. PICP increased after 1 week [9, 43], whereas osteocalcin and bone alkaline phosphatase isoenzyme augmented after 3 months of GH treatment [9, 44, 47]. The changes of PICP levels during the first week [9] and after 3 months [30], as well as the changes of bone alkaline phosphatase isoenzyme after 3 months [44, 47] of therapy were positively related to the changes in growth velocity at 12 months. This provides evidence that serum PICP and bone alkaline phosphatase isoenzyme concentrations in patients with GH deficiency may be useful markers to predict growth velocity during long-term GH treatment. We did not find such a relationship for osteocalcin [9], but this evidence was reported by other authors [40–42]. Moreover, it has been demonstrated that GH had a dose-dependent effect on PICP, osteocalcin, and bone alkaline phosphatase isoenzyme. Serum PICP levels increased more at a GH dose of 4 IU/day compared with 2 IU/day, whereas no further increase was found at a GH dose of 6 IU/day. Such data suggest that a GH replacement dose of 4 IU/day near maximally stimulated the type I collagen synthesis [48]. As regards osteocalcin and bone alkaline phosphatase isoenzyme, it has been reported that a GH dose of 2 IU/day was inadequate to maintain normal levels of these bone markers [49]. In addition, GH-treatment was able to normalize the altered nighttime osteocalcin pattern found in GH-deficient children [49] and in children with idiopathic short stature [50]. Serum PIIINP values at 5 weeks [51] or after 3 months [45] of GH treatment were positively related to growth velocity at 12 months; thus, it may be an additional biochemical marker to predict growth velocity during the hormonal therapy. Moreover, GH treatment increased circadian PIIINP concentrations in children with idiopathic short stature [52].

Thus, the measurement of bone- and growth-related biochemical markers may be helpful to study bone turnover and somatic growth, in response to GH treatment.

Effects of GH on Bone Density

In addition to the effects on linear growth, GH has a critical role in the attainment of bone mass. GH-deficient children have osteopenia partly because of delayed bone maturation [9, 53, 54]; thus, GH deficiency itself is implicated with the development of osteopenia. It has been also demonstrated that acquired GH deficiency before the completion of puberty results in severe osteopenia in adult life [55]. Recent studies have demonstrated that the timing of peak bone mass occurs during adolescence, within the age of 18 years for both cortical and trabecular bone, even though some differences exist between the skeletal sites and sexes [56–58]. In females, lumbar peak bone mass occurs at 14–15 years [56, 57] whereas radial peak bone mass occurs at 18–20 years [58]. In males, lumbar peak bone mass is delayed compared with females, as it occurs at about 18 years; however, no delay is observed in relation to pubertal stage [56, 57]. Anyway, in normal circumstances, the major amount of peak bone mass is achieved before 18–20 years in both sexes [56–58]. After this age, the increase in bone mass is very small, as demonstrated in the forearm [58]. The decline of bone density, in particular in the proximal end of the femur and in the spine, occurs immediately after peak bone mass has been attained [57, 58].

GH-deficient adults with childhood- and adult-onset pituitary failure have reduced bone density, with an increased risk of osteoporosis [59–64]. Other studies [55, 65, 66] did not confirm these data. Anyway, approximately 17% of GH-deficient adults have vertebral fractures, compared with 8–10% of the normal population [67]. In this regard, reduced bone density in GH-deficient adults with childhood-onset not only reflects the consequences of a prolonged GH deficiency during childhood, but it may also be related to the absence of GH during a period of early childhood when peak bone mass is perhaps not yet reached. On the contrary, Ogle et al. [68] in GH-deficient adolescents showed increased values of bone density in total body, in spine, and in femur after discontinuation of GH therapy. However, this increment in bone density may be complicated by an effect of puberty on bone mass, at least in some patients. In the opinion of the authors, it is more likely that the 12-month period of follow-up was too short to show a decline in bone density.

A significant improvement in bone density has been demonstrated in GH-deficient prepubertal children during GH treatment; after 12 months of GH, about 50% of children normalized their bone density, measured by single-photon absorptiometry at the distal one-third of the nondominant radius, and calculated for both bone and statural ages [9]. In particular, GH treatment determined an increase in bone density greater than expected by the increment in height [9]. After 6 months of GH, Zamboni et al. [54] showed a significant increase in bone density, measured at the ultradistal radius by using dual-photon absorptiometry, even if the patients failed to normalize their bone density at this time. On the contrary, after 1 year of GH treatment, Shore et al. [53] did not observe any variation in bone density measured at nondominant radius, ulna, and humerus, by using single-photon absorptiometry. Anyway, even though the reduced bone density did not recover under GH therapy, osteopenia did not become worse during treatment when growth velocity doubled. However, the dose of GH administered by Shore et al. [53] was lower than that we [9] and Zamboni et al. [54] used. According to Shore et al. [53], Lu et al. [69] did not find any significant improvement of bone density in spine, measured by dual x-ray absorptiometry, after 1 year of GH treatment in patients with GH deficiency; whereas short normal children showed a significant increase in bone density after the same time of treatment. Although the results are not univocal, these data seem to indicate that GH treatment positively influences bone mass. The positive relationship between bone density and $TMPO_4$/GFR ratio we found during GH treatment (bone density corrected for statural age $r = 0.91$, $p < 0.001$; bone density corrected for bone age $r = 0.90$, $p < 0.001$) seems to indicate that a higher phosphate retention may be a factor involved in improving bone density in children with GH deficiency [9].

GH treatment in GH-deficient adults increased bone density [59, 61, 70] and bone turnover [40, 63, 67, 70, 71], but long-term studies are needed to verify these effects on adult bone. The question is if the reduced bone density in GH-deficient adults derives from an insufficient attainment of peak bone mass during childhood or if it is a consequence of premature bone loss during adult life. Kaufman et al. [60] in the longitudinal part of their study did not find a decline of bone density after 6–28 months without GH treatment, suggesting that the osteopenia in GH-deficient adults may be rather the consequence of a deficient build-up of bone mass during childhood and adolescence. As a confirmation, no significant correlation between the duration of GH deficiency and bone density was found [60, 64]. On the contrary, a negative correlation between these two parameters was reported by Hyer et al. [55]. Kaufman's data [60] together with our's [9], that optimal GH replacement seems to be able to normalize bone density in some GH-deficient children, suggest that a decreased peak bone mass is likely the primary cause of reduced bone density in GH-deficient young adults. Reduced peak bone mass may be the consequence of suboptimal GH treatment as inadequate GH doses, and duration and/or timing of GH treatment [60, 64]. However, an increased rate of bone loss may also occur after the GH treatment is stopped, but data on this matter are lacking. On the other hand, GH-deficient adults may have experienced a period of sex hormone deficiency or a delayed puberty before starting GH treatment. So, a brief period of hypogonadism may be also associated with osteopenia in patients with GH deficiency [55]. These findings, together with the observation that adult men with a history of constitutionally delayed puberty have decreased radial and spinal bone density [72], and that boys with constitutionally delayed puberty and reduced bone density did not reach normal bone mass in adulthood [73], suggest that the timing of puberty is a major determinant in the attainment of peak bone mass, as previously appeared from a study of Krabbe et al. [74]. Therefore, GH replacement must be started as early as possible in GH-deficient children to improve longitudinal growth and to sustain bone density, even though we did not know the optimal GH doses to reach the adequate bone mass in these patients. In GH-deficient adults, GH treatment may be indicated to improve bone density and to prevent the premature development of osteoporosis, but additional data are needed. Thus, a longitudinal follow-up of bone density is advised during GH treatment and after the cessation of linear growth.

In vitro Effects of GH on Chondrocytes and Osteoblast-Like Cells

There are no doubts that GH exerts relevant effects either on growing bone or on mature bone [1, 3, 75]; whether the hormone influences the biological activity of bone cells directly or, as it happens in other tissues, through IGF-I, it remains for certain aspects to be elucidated. As regards the growing bone, and in particular the growth plate, studies revealed that these two peptides have synergistic effects at that level. Long bone growth is stimulated directly and in a dose-dependent manner [1] by GH, whose presence is required for chondrocyte differentiation from precursors. The differentiation process makes chondrocytes sensitive to IGF-I, the local production of which is also increased by GH, that in turn stimulates in an autocrine/paracrine fashion the clonal expansion of differentiating chondrocytes [1]. At this stage, IGF-I, but not GH, seems relevant to chondrocyte proliferation as demonstrated by the failure of GH and IGF-I to stimulate their proliferation when cells are incubated with an antibody specifically reacting to IGF-I receptor [76]. These data indirectly agree with our own, demonstrating the absence of an additive effect on chondrocyte proliferation when GH and IGF-I are added in combination to cultures [77]. Recent evidence has shown that human articular chondrocytes may produce and are sensitive to several cytokines, as interleukin (IL)-1, IL-6, IL-8 and transforming growth factor-β (TGF-β) [78–81]. Furthermore, studies on fracture repair, a process that requires chondrogenesis and endochondral ossification in its latest phases, demonstrated that growth factors, including TGF-β and acidic fibroblast growth factors, are central regulators of cellular proliferation, differentiation and extracellular matrix synthesis [82]. By using the cDNA PCR technique we found the transcripts of IL-1β, IL-6, IL-8 and TGF-β in human chondrocytes of the growing cartilage of the iliac crest, cultured in vitro. We observed that the addition to the cultures of varying concentrations of GH (from 10 to 40 ng/ml) or IGF-I (from 20 to 80 mIU/ml) significantly and dose-dependently increased the levels of IL-6 but not those of the remaining cytokines in the supernatants of the cultures. Furthermore, IGF-I showed a higher potency in doing that. The exact role of IL-6 in cartilage is still obscure; recent studies indicated that this multifunctional cytokine is critical to the formation of osteoclasts from less differentiated cells [83] and that it induces the synthesis of a tissue inhibitor of cartilage-degrading enzymes, the metalloproteinases [84], suggesting that chondrocytes may influence cartilage degradation and bone formation under particular circumstances.

As regards the effects of GH and in particular its interaction with IGF-I in mature bone, less it is understood. In animal models, GH treatment increases in vitro the number and the function of osteoblasts [85], but whether or not these are direct effects, not mediated by IGF-I it is not yet secured. We found that both GH and IGF-I dose-dependently and significantly stimulated the proliferation of the human osteosarcoma-derived osteoblast-like cell line SaOS-2; anyway, we failed to observe an additive effect on cell proliferation when cultures were incubated with both peptides; this indirectly confirmed that GH regulated osteoblast-like cell proliferation via the paracrine/autocrine production of IGF-I [77]. Furthermore, there is evidence that GH stimulates the production of IGF-I in rat osteoblast-like cells, and IGF-I itself may induce the synthesis of bone matrix [18] and osteocalcin [86]. We found that in our culture system both GH and IGF-I were able to significantly increase the levels of this bone protein in the supernatant [unpubl. data]. Again, the GH-induced action on osteocalcin may be due to the induction of locally produced IGF-I.

The exact role of the various growth factors and cytokines produced by cells during bone growth is at present largely unknown; therefore, the events observed in vitro do not necessarily reflect what happens in vivo, when different cells, in a different stage of maturation, that are growing on a particular matrix, receive several, sometimes contrasting signals. This picture is further complicated by the possible modulatory action of various hormones. It may be that in certain conditions a different GH secretory status may influence the biological activity of these cells.

References

1 Isaksson OGP, Lindahl A, Nilsson A, Isgaard J: Mechanism of the stimulatory effect of growth hormone on longitudinal bone growth. Endocr Rev 1987;8:426–438.

2 Ogle GD, Rosenberg AR, Kainer G: Renal effects of growth hormone. II. Electrolyte homeostasis and body composition. Pediatr Nephrol 1992;6:483–489.

3 Inzucchi SE, Robbins RJ: Effects of growth hormone on human bone biology. J Clin Endocrinol Metab 1994;79:691–694.

4 Saggese G, Baroncelli GI, Bertelloni S, Cinquanta L, Di Nero G: Twenty-four-hour osteocalcin, carboxyterminal propeptide of type I procollagen, and aminoterminal propeptide of type III procollagen rhythms in normal and growth-retarded children. Pediatr Res 1994;35:409–415.

5 Wuster C: Growth hormone and bone metabolism. Acta Endocrinol (Copenh) 1993;128(suppl 2):14–18.

6 Chipman JJ, Zerwekh J, Nicar M, Marks J, Pak CJC: Effect of growth hormone administration: Reciprocal changes in serum 1α,25-dihydroxyvitamin D and intestinal calcium absorption. J Clin Endocrinol Metab 1980;51:321–324.

7 Henneman PH, Forber AP, Modawer M, Dempsey EF, Carrol EL: Effects of human growth hormone in man. J Clin Invest 1960;39:1223–1238.

8 Bruns MEH, Vollmer SS, Bruns DE, Overpeck JG: Human growth hormone increases intestinal vitamin D dependent calcium-binding protein in hypophysectomized rats. Endocrinology 1983;113:1387–1392.

9 Saggese G, Baroncelli GI, Bertelloni S, Cinquanta L, Di Nero G: Effects of long-term treatment with growth hormone on bone and mineral metabolism in children with growth hormone deficiency. J Pediatr 1993;122:37–45.

10 Gertner JM, Horst RL, Broadus AE, Rasmussen H, Genel M: Parathyroid function and vitamin D metabolism during human growth hormone replacement. J Clin Endocrinol Metab 1979;49:185–188.

11 Nishiyama S, Ikuta M, Nakamura T, Tomoeda S, Matsuda I: Renal handling of phosphate can predict height velocity during growth hormone therapy for short children. J Clin Endocrinol Metab 1992;74:906–909.

12 Gertner JM, Tamborlane WV, Hintz RL, Horst RL, Genel M: The effects on mineral metabolism of overnight growth hormone infusion in growth hormone deficiency. J Clin Endocrinol Metab 1981;53:818–822.

13 Caversazio J, Montessuit C, Bonjour JP: Stimulatory effect of insulin-like growth factor-I on renal Pi transport and plasma 1,25-dihydroxyvitamin D_3. Endocrinology 1990;127:453–459.

14 Harbison MD, Gertner JM: Permissive action of growth hormone on the renal response to dietary phosphorus deprivation. J Clin Endocrinol Metab 1990;70:1035–1040.

15 Caversazio J, Bonjour JP, Fleisch H: Adaption of tubular phosphate transport: Relation between phosphate requirement as influenced by growth and supply. Adv Exp Med Biol 1989;128:107–111.

16 Kruse K, Kracht U, Gopfert G: Renal threshold phosphate concentration ($TMPO_4$/GFR). Arch Dis Child 1982;57:217–223.

17 Saggese G, Bertelloni S, Baroncelli GI, Bottone E: Livelli dei metaboliti della vitamina D in età pediatrica. Dati italiani. Riv Ital Pediatr 1986;12:560–564.

18 Bouillon R: Growth hormone and bone. Horm Res 1991;36(suppl):49–55.

19 Lucaszkiewcz J, Lorenc R, Romer T, Garabedian M, Balsan S, Jones J: Growth hormone-induced catch-up growth and serum $1,25(OH)_2D$ concentrations in pituitary dwarfs; in Norman AW (ed): Vitamin D. Chemical, Biochemical and Clinical Update. Berlin, de Gruyter, 1985, pp 1109–1110.

20 Gray RW, Garthwaite TL: Activation of renal $1,25(OH)_2D$ synthesis by phosphate deprivation: Evidence for a role for growth hormone. Endocrinology 1985;116:189–193.

21 Halloran BP, Spencer EM: Dietary phosphorus and 1,25-dihydroxyvitamin D metabolism: Influence of insulin-like growth factor-I. Endocrinology 1988;123:1225–1229.
22 Burstein S, Chen IW, Tsang RC: Effects of growth hormone replacement therapy on 1,25-dihydroxyvitamin D and calcium metabolism. J Clin Endocrinol Metab 1983;56:1246–1251.
23 Spanos E, Barret D, MacIntyre I, Pike JW, Safilian EF, Haussler MR: Effect of growth hormone on vitamin D metabolism. Nature 1978;273:246–247.
24 Marcus R, Butterfield G, Holloway L, Gilliland L, Baylink DJ, Hintz RL, Sherman BM: Effects of short-term administration of recombinant human growth hormone to elderly people. J Clin Endocrinol Metab 1990;70:519–527.
25 Lancer SR, Bowser EN, Hargis GK, Williams GA: The effect of growth hormone on parathyroid function in rats. Endocrinology 1976;98: 1289–1293.
26 Van der Veen EA, Netelenbos JC: Growth hormone (replacement) therapy in adults: Bone and calcium metabolism. Hormone Res 1990; 33(suppl 4):65–68.
27 Kruse K: Effect of biosynthetic human growth hormone on calcium and bone metabolism; in Preece MA (ed): International Symposium on Growth Disorders. Princeton, Excerpta Medica, 1987, pp 54–67.
28 Johansen JS, Giwercman A, Hartwell D, Nielsen CT, Price PA, Christiansen C, Skakkebaek NE: Serum bone Gla protein as a marker of bone growth in children and adolescents: Correlation with age, height, serum insulin-like growth factor I, and serum testosterone. J Clin Endocrinol Metab 1988;67:273–278.
29 Saggese G, Baroncelli GI, Bertelloni S, Buggiani B: Livelli normali di osteocalcina in età pediatrica. Comparazione con la mineralizzazione ossea e con le variazioni fisiologiche dei livelli di 1,25-diidrossivitamina D. Riv Ital Pediatr 1989;15:109–112.
30 Trivedi P, Risteli J, Risteli L, Hindmarsh PC, Brook CGD, Mowat AP: Serum concentrations of the type I and III procollagen propeptides as biochemical markers of growth velocity in healthy infants and children and in children with growth disorders. Pediatr Res 1991;30: 276–280.
31 Delemarre-van de Waal HA, Martens F, Odnik RJH: Type I procollagen levels in relation to growth, growth hormone, SmC and sex steroids throughout normal puberty (abstract). Horm Res 1991;35(suppl 2):41.
32 Saggese G, Bertelloni S, Baroncelli GI, Di Nero G: Serum levels of carboxyterminal propeptide of type I procollagen in healthy children from 1st year of life to adulthood and in metabolic bone diseases. Eur J Pediatr 1992;151:764–768.
33 Hertel NT, Stoltenberg M, Muller J, Lorenzen I, Skakkebaek NE: Procollagen I and III in children and adolescents: Correlation to linear growth and pubertal maturation (abstract). Horm Res 1992;37(suppl 4):57.
34 Parfitt AM, Simon LS, Villanueva AR, Krane SM: Procollagen type I carboxyterminal extension peptide in serum as a marker of collagen biosynthesis in bone. Correlation with iliac bone formation rates and comparison with total alkaline phosphatase. J Bone Miner Res 1987;2:427–436.
35 Deftos LJ: Bone protein and peptide assays in the diagnosis and management of skeletal disease. Clin Chem 1991;37:1143–1148.
36 Charles P, Hasling C, Risteli L, Risteli J, Mosekilde L, Eriksen EF: Assessment of bone formation by biochemical markers in metabolic bone disease: Separation between osteoblastic activity at the cell and tissue level. Calcif Tissue Int 1992;51:406–411.
37 Risteli J, Risteli L: Growth and collagen. Curr Med Lit – Growth and Growth Factors 1989;4: 159–164.
38 Prockop DJ, Kkivirikko KJI, Tuderman L, Guzman A: The biosynthesis of collagen and its disorders. N Engl J Med 1979:301:13–23, 77–85.
39 Delmas PD, Chatelain P, Malaval L, Bonne G: Serum bone Gla protein in growth hormone-deficient children. J Bone Miner Res 1986;1: 333–338.
40 Johansen JS, Jensen SB, Riis BJ, Rasmussen L, Zachman M, Christiansen C: Serum bone Gla protein: A potential marker of growth hormone (GH) deficiency and the response to GH therapy. J Clin Endocrinol Metab 1990;71:122–126.
41 Sartorio A, Coni A, Guzzaloni G, Faglia G: Serum osteocalcin levels in patients with GH deficiency before and during GH treatment. Acta Paediatr Scand 1991;80:100–102.
42 Kanzaki S, Hosoda K, Moriwake T, Tanaka H, Kubo T, Inoue M, Higuchi J, Yamaji T, Seino Y: Serum propetide and intact molecular osteocalcin in normal children and children with growth hormone (GH) deficiency: A potential marker of bone growth and response to GH therapy. J Clin Endocrinol Metab 1992;75: 1104–1109.
43 Carey DE, Goldberg B, Ratzan SK, Rubin KR, Rowe DW: Radioimmunoassay for type I procollagen in growth hormone-deficient children before and during treatment with growth hormone. Pediatr Res 1984;19:8–11.
44 Crofton PM, Stirling HF, Kelnar CJH: An increase in bone alkaline phosphatase can be used to predict responses to growth-promoting treatments. Acta Paediatr Scand 1991; 379(suppl):181.
45 Trivedi P, Hindmarsh PC, Risteli J, Risteli L, Mowat AP, Brook CGD: Growth velocity, growth hormone therapy, and serum concentrations of the amino-terminal propeptide of type III procollagen. J Pediatr 1989;114:225–230.
46 Danne T, Gruters A, Schuppan D, Quantas N, Enders I, Weber B: Relationship of procollagen type II propeptide-related antigens in serum to somatic growth in healthy children and patients with growth disorders. J Pediatr 1989; 114:257–260.
47 Bertelloni S, Baroncelli GI, Cappuccio A, Pardini E, Innocenti B, Saggese G: Bone alkaline phosphatase: A predictive marker of growth response in 'nonclassic' growth hormone-deficient children during growth hormone treatment (abstract). Proc 15th Int Symp on Growth and Growth Disorders, Florence, April 2–3, 1993.
48 Jensen LT, Jorgensen JOL, Risteli J, Christiansen JS, Lorenzen IB: Type I and III procollagen propeptides in growth hormone-deficient patients: Effects of increasing doses of GH. Acta Endocrinol (Copenh) 1991;124:278–282.
49 Nielsen HK, Jorgensen JOL, Brixen K, Christiansen JS: Serum osteocalcin and bone isoenzyme alkaline phosphatase in growth hormone-deficient patients: Dose-response studies with biosynthetic human GH. Calcif Tissue Int 1991;48:82–87.
50 Markowitz ME, Dimartino-Nardi J, Gasparini F, Fishman K, Rosen JF, Saenger P: Effects of growth hormone therapy on circadian osteocalcin rhythms in idiopathic short stature. J Clin Endocrinol Metab 1989;69:420–425.
51 Tapanainen P, Risteli L, Knip M, Kaar ML, Risteli J: Serum aminoterminal propeptide of type III procollagen: A potential predictor of the response to growth hormone therapy. J Clin Endocrinol Metab 1988;67:1244–1249.
52 Markowitz ME, Saenger PH, Gasparini FJ, Fishman KP: 24-Hour procollagen III levels during growth hormone treatment of children with idiopathic short stature (abstract). Pediatr Res 1989;25:88A.
53 Shore RM, Chesney RW, Mazess RB, Rose PG, Bargman GJ: Bone mineral status in growth hormone deficiency. J Pediatr 1980;96: 393–396.
54 Zamboni G, Antoniazzi F, Radetti G, Musumeci C, Tat L: Effect of two different regimens of recombinant human growth hormone therapy on the bone mineral density of patients with growth hormone deficiency. J Pediatr 1991; 119:483–485.
55 Hyer SL, Rodin DA, Tobias JH, Leiper A, Nussey SS: Growth hormone deficiency during puberty reduces adult bone mineral density. Arch Dis Child 1992;67:1472–1474.
56 Bonjour J-PH, Theintz G, Buchs B, Slosman D, Rizzoli R: Critical years and stages of puberty for spinal and femoral bone mass accumulation during adolescence. J Clin Endocrinol Metab 1991;73:555–563.
57 Theintz G, Buchs B, Rizzoli R, Slosman D, Clavien H, Sizonenko PC, Bonjour J-PH: Longitudinal monitoring of bone mass accumulation in healthy adolescents: Evidence for a marked reduction after 16 years of age at the levels of lumbar spine and femoral neck in female subjects. J Clin Endocrinol Metab 1992; 75:1060–1065.
58 Matkovic V, Jelic T, Wardlaw GM, Llich JZ, Goel PK, Wright JK, Andon MB, Smith KT, Heaney RP: Timing of peak bone mass in Caucasian females and its implication for the prevention of osteoporosis. Inference from a cross-sectional model. J Clin Invest 1994;93:799–808.

59 Degerblad M, Elgindy N, Hall K, Sjoberg HE, Thoren M: Potent effect of recombinant growth hormone on bone mineral density and body composition in adults with panhypopituitarism. Acta Endocrinol (Copenh) 1992;126: 387–393.
60 Kaufman J-M, Taelman P, Vermeulen A, Vandeweghe M: Bone mineral status in growth hormone-deficient males with isolated and multiple pituitary deficiencies of childhood onset. J Clin Endocrinol Metab 1992;74:118–123.
61 O'Halloran DJ, Tsatsoulis A, Whitehouse RW, Holmes SJ, Adams JE, Shalet SM: Increased bone density after recombinant human growth hormone (GH) therapy in adults with isolated GH deficiency. J Clin Endocrinol Metab 1993; 76:1344–1348.
62 Rosen T, Hansson T, Granhed H, Szucs J, Bengtsson BA: Reduced bone mineral content in adult patients with growth hormone deficiency. Acta Endocrinol (Copenh) 1993;129: 201–206.
63 Thoren M, Soop M, Degerblad M, Saaf M: Preliminary study of the effects of growth hormone substitution therapy on bone mineral density and serum osteocalcin levels in adults with growth hormone deficiency. Acta Endocrinol (Copenh) 1993;128(suppl 2):41–43.
64 De Boer H, Blok GJ, Van Lingen A, Teule GJJ, Lips P, Van Der Veen EA: Consequences of childhood-onset growth hormone deficiency for adult bone mass. J Bone Miner Res 1994;9: 1319–1326.
65 Johansson AG, Burman P, Westermark K, Ljunghall S: The bone mineral density in acquired growth hormone deficiency correlates with circulating levels of insulin-like growth factor I. J Intern Med 1992;232:447–452.
66 Kaji H, Fukase M, Abe H, Chihara K: Evaluation of bone mineral density in adults with acquired growth hormone deficiency (abstract). Acta Endocrinol (Copenh) 1993; 128(suppl 2):61.
67 Wuster C, Ljunghall S: Workshop report: The effects of growth hormone and growth hormone deficiency on bone tissues. Acta Endocrinol (Copenh) 1993;128(suppl 2):62.
68 Ogle GD, Moore B, Lu PW, Craighead A, Briody JN, Cowell CT: Changes in body composition and bone density after discontinuation of growth hormone therapy in adolescence: An interim report. Acta Paediatr Scand 1994; 83(suppl 399):3–7.
69 Lu PW, Cowell CT, Moore B, Craighead A, Howman-Giles R, Briody J: Spine bone mineral density and growth hormone treatment. Acta Paediatr Scand 1992;383(suppl):95.
70 Vandeweghe M, Taelman P, Kaufman JM: Short- and long-term effects of growth hormone treatment on bone turnover and bone mineral content in adult growth hormone-deficient males. Clin Endocrinol 1993;39:409–415.
71 Schlemmer A, Johansen JS, Pedersen SA, Jorgensen JOL, Hassager C, Christiansen C: The effect of growth hormone (GH) therapy on urinary pyridinoline cross-links in GH-deficient adults. Clin Endocrinol 1991;35:471–476.
72 Finkelstein JS, Neer RM, Biller BMK, Crawford JD, Klibanski A: Osteopenia in men with a history of delayed puberty. N Engl J Med 1992;326:600–604.
73 Bertelloni S, Baroncelli GI, Saggese G: Osteopenia in constitutionally delayed puberty: A longitudinal study (abstract). Proc 75th Annu Meet of the Endocrine Society, Las Vegas, Nev., June 9–12, 1993, p 308.
74 Krabbe S, Christiansen C, Rodbro P, Transbol I: Effect of puberty on rates on bone growth and mineralisation. With observations in male delayed puberty. Arch Dis Child 1979;54:950–953.
75 Lindahl A, Nilsson A, Isaksson OGP: Effects of growth hormone and isulin-like growth factor-I on colony formation of rabbit epiphyseal chondrocytes at different stages of maturation. J Endocrinol 1987;115:263–271.
76 Brenner RE, Nerlich A, Heinzee E, Vetter U, Teller W: Different regulation of clonal growth by transforming factor-β_1 in human fetal articular and costal chondrocytes. Pediatr Res 1993;33:390–393.
77 Saggese G, Federico G, Cinquanta L: In vitro effects of growth hormone and other hormones on chondrocytes and osteoblast-like cells. Acta Paediatr Scand 1993;391(suppl):55–59.
78 Guerne PA, Carson DA, Lotz M: IL-6 production by human articular chondrocytes. Modulation of its synthesis by cytokines, growth factors, and hormones in vitro. J Immunol 1990; 144:499–505.
79 Tiku K, Thakker-Varia S, Ramachandrula A, Tiku ML: Articular chondrocytes secrete IL-1, and have IL-1 inhibitory activity. Cell Immunol 1992;140:1–20.
80 Lotz M, Terkeltaub R, Villiger PM: Cartilage and joint inflammation. Regulation of IL-8 expression by human articular chondrocytes. J Immunol 1992;148:466–473.
81 Chu CQ, Field M, Ilards S, Abney E, Feldmann M, Maini RN: Detection of cytokines at the cartilage/pannus junction in patients with rheumatoid arthritis: Implications for the role of cytokines in cartilage destruction and repair. Br J Rheumatol 1992;31:653–661.
82 Bolander ME: Regulation of fracture repair by growth factors. Proc Soc Exp Biol Med 1992; 200:165–170.
83 Kurihara N, Civin C, Rooditan GD: Osteotropic factor responsiveness of highly purified populations of early and late precursors for human multinucleated cells expressing the osteoclast phenotype. J Bone Miner Res 1991;6: 257–261.
84 Lotz M, Guerne PA: Interleukin-6 induces the synthesis of tissue inhibitor of metalloproteinases-1/erythroid potentiating activity. J Biol Chem 1991;266:2017–2020.
85 Slootweg MC, Van Bool-Offers SC, Herrmann-Erlee MpM, Duursma SA: Direct stimulatory effect of growth hormone on DNA synthesis of fetal chicken osteoblasts in culture. Acta Endocrinol (Copenh) 1988;118:294–300.
86 Canalis E, Lian JB: Effects of bone-associated growth factors on DNA, collagen and osteocalcin synthesis in cultured fetal rat calvariae. Bone 1988;9:243–246.

Horm Res 1995;44(suppl 3):64–72

Anders Juul[a]
Jens Otto L. Jørgensen[b]
Jens S. Christiansen[b]
Jørn Müller[a]
Niels E. Skakkebæk[a]

[a] Department of Growth and Reproduction, University of Copenhagen;
[b] Department of Internal Medicine (Diabetes and Endocrinology), University of Aarhus, Denmark

Metabolic Effects of GH: A Rationale for Continued GH Treatment of GH-Deficient Adults after Cessation of Linear Growth

Key Words
GH treatment
GH deficiency
Adults
Diagnosis

Abstract
GH deficiency in children has been treated for decades with GH in order to normalize growth and final height. After cessation of linear growth GH treatment has traditionally been stopped. However, adults with GH deficiency have increased mortality due to cardiovascular disease and share several clinical signs, such as reduced muscle strength, reduced exercise capacity, reduced sweating ability and thermoregulation, subnormal kidney function, reduced myocardial function, reduced energy expenditure, abnormal thyroid hormone metabolism, impaired psychosocial well-being, decreased lean body mass and decreased bone mineral content. Several studies have shown unanimously that GH replacement therapy in GH-deficient adults have improved all the above-reported clinical symptoms. We conclude that GH deficiency in adults is associated with a number of subnormal features. GH treatment improves (normalizes) these symptoms and we therefore believe that GH replacement therapy of a GH-deficient individual should be continued after cessation of linear growth. However, treatment of such patients should be carefully monitored since data from follow-up studies are still lacking.

Introduction

It is well recognized that GH treatment promotes linear growth in short children with GH deficiency. In addition, GH-deficient children have an abnormal body composition (reduced lean body mass and excessive fat mass) that is normalised by GH treatment. Most of these results have been accomplished by the use of extracted pituitary hGH. With the recent advent of biosynthetic human GH wider indications for treatment with GH are now possible.

The traditional view that GH is only of importance for linear growth has led to a lack of interest in the possible continuation of GH replacement therapy beyond the cessation of linear growth in the adolescent period. In addition, it has not been well examined what the consequences of withdrawal of GH replacement in this critical period might be. However, during the last 5-year period, a substantial amount of data has shown that a number of abnormal features exist in GH-deficient adults and that these can be normalized if GH replacement therapy is either reinstituted or introduced.

Normal ageing is associated with a gradual decline in GH secretion [1] accompanied by a reduced bone mineral content, lean body mass and increased fat mass, a phenomenon referred to as the somatopause. In this respect,

Anders Juul
Department of Growth and Reproduction
University of Copenhagen
Blegdamsvej 9
DK–2100 Copenhagen (Denmark)

0301–0163/95/0449–0064
$8.00/0

there are several resemblances between physiological ageing and adult GH deficiency, and GH treatment has actually been attempted in healthy elderly people [2, 3] with rejuvenating effects. However, these aspects of GH treatment have to be clearly segregated from treatment of adults with verified 'true' GH deficiency. In this paper, we will describe only the typical features of adult patients with 'true' GH deficiency and the potential effects of GH treatment.

Diagnosis of GH Deficiency

Obviously, some of the potential candidates for adult GH replacement therapy have been diagnosed in childhood due to growth deficit and have subsequently been treated with GH. When diagnosing adults with GH deficiency or confirming the diagnosis in adult patients with GH deficiency of childhood onset, we have to rely on clinical data (specific features of GH deficiency), provocative tests (at least two separate tests; pharmacological and/or physiological) and laboratory measurement (serum IGF-I, IGFBP-3 and urinary GH excretion), as auxological data do not support the diagnosis in adults. However, the GH-IGF hormone axis offers many challenges (laboratory-related, biological as well as obtaining valid normative data). This implicates some pitfalls, which have to be dealt with in the future.

In the following we shall briefly discuss these problems.

Lack of Valid Reference Data

(1) *Spontaneous GH secretion.* Normal ageing is associated with a gradual decline in integrated 24-hour GH secretion [1] and very few laboratories have a substantial reference material on healthy adults of different age groups.

(2) *GH provocative test: definition of specific cut-off value for each laboratory.* Due to the laboratory difficulties when measuring GH and to differences in standardization (22 kD recombinant GH form or pituitary-derived GH) each laboratory should construct its own cut-off level [4]. In addition, many different physiological and pharmacological agents (heat exposure, insulin, arginin, clonidine, *L*-Dopa, GHRH, etc.) have been used and the GH response may vary according to the stimulus [5]. Finally, GH responsivity to a provocative test is strongly dependent on other factors, such as nutritional status and sex steroids, and the GH response in women is therefore dependent on the time of menstrual cycle [6].

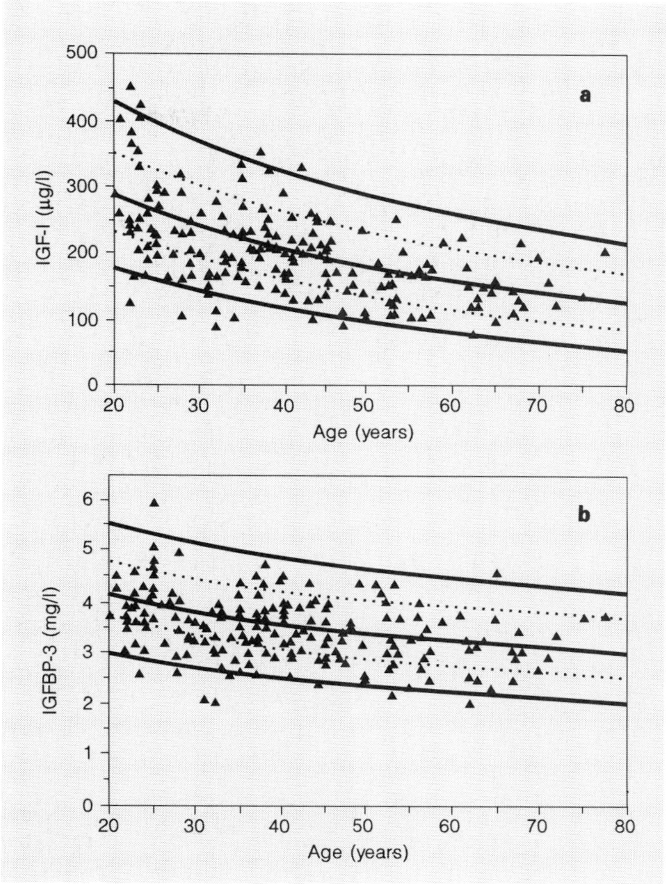

Fig. 1. Serum levels of IGF-I and IGFBP-3 in 198 healthy individuals decline significantly with age ($p < 0.0001$). The solid lines represent mean and 95% prediction interval; the dotted lines represent ± 1 SD. Data are derived from Juul et al. [12], with permission.

(3) *Serum IGF-I and IGFBP-3.* IGF-I and IGFBP-3 have been reported as excellent diagnostic markers for GH deficiency in children by most authors [7, 9], but not all [10], as relatively well-defined normative data exist for children. However, few valid reference materials exist for IGF-I and IGFBP-3 in adults and the diagnostic value of these parameters in adults has not been evaluated thoroughly at present. Serum levels of IGF-I and IGFBP-3 correlate with spontaneous GH secretion in children [11] and exhibit, like integrated serum GH concentrations, a gradual decline with increasing age in healthy adults [12] (fig. 1). This implies that a hypothetical IGF-I concentration of 150 µg/l is below the normal range for a 20-year-old patient but in the middle range of the reference material if the patient is 70 years old. However, serum concentrations of IGF-I and IGFBP-3 are dependent on other factors,

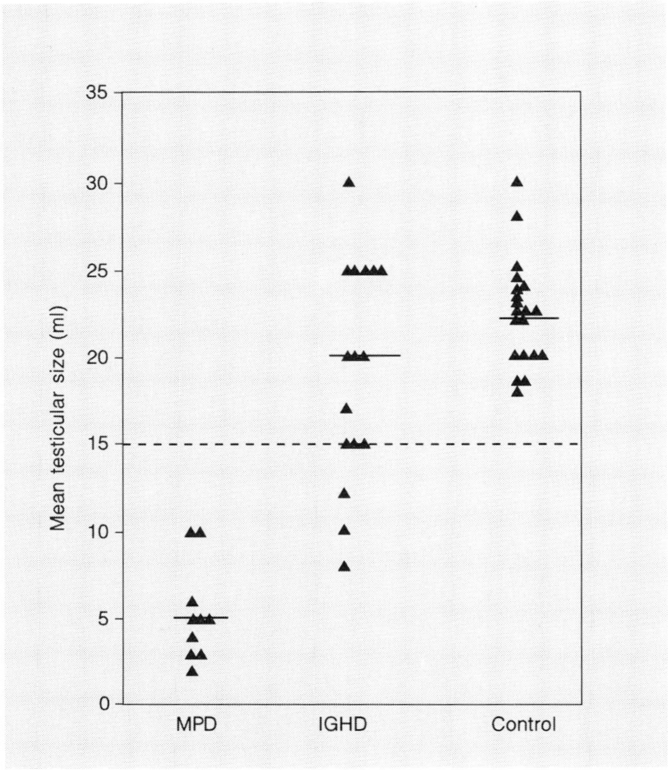

Fig. 2. Average testicular volume in patients with multiple pituitary deficiency (MPD), isolated growth hormone deficiency (IGHD) and normal volunteers [88]. Data are derived from Pedersen et al. [49].

such as nutrition [13, 14], hepatic function [15, 16], renal function [17, 18], diabetes [19] and thyroid disease [20]. These findings emphasize the importance of a large reference material and examining for other possible diseases, when utilising serum concentrations of IGF-I and IGFBP-3 as diagnostic tools for GH deficiency in adults.

Laboratory Problems

(1) Heterogeneity of circulating GH forms (22 kD, 20 kD, acidic and dimeric forms) affects the affinity of different antibodies (mono-/polyclonal) to a varying extent and makes comparisons between different laboratories difficult.

(2) Interference of IGF binding proteins in IGF-I assays makes the time-consuming and laborious extraction of serum prior to measurement mandatory.

(3) Lack of international standardization of IGFBP-3 measurements.

Based on these comments, we therefore propose that a definition of the adult GH deficiency syndrome should at least include: (1) insufficient GH response to a classical stimulation test (taking into account age, stage in menstrual cycle, and assay specific cut-off values), *and* (2) serum IGF-I and IGFBP-3 levels below the normal range for age (large reference material required), *and* (3) GH deficiency since childhood, *or* (4) if GH deficiency is acquired in adulthood, then additional deficiency of at least one tropic hormone (ACTH, LH/FSH, TSH) or substantial evidence of pituitary disease (intracranial tumor, irradiation).

Features of the Adult GH Deficiency Syndrome

Body Composition

GH-deficient adults have a decreased lean body mass and a decreased muscle/fat ratio as demonstrated by various methods: cross-sectional CT scan of the thigh [21], bioelectrical impedance [22, 23] and total body potassium assessment [22, 24]. Also a decreased extracellular fluid volume has been reported in these patients [24]. Furthermore, a reduced energy expenditure has been described [25] possibly due to the decreased lean body mass.

Muscular Performance

Reduced exercise capacity [26, 27] as well as reduced muscle strength [26] was demonstrated in GH-deficient adults, although histological studies of skeletal muscles showed no signs of myopathy [28].

Cardiovascular System

GH-deficient patients have myocardial dysfunction [29–31], 'atherogenic' lipoprotein profile [32, 33], abdominal adiposity (a known cardiovascular risk factor), and perhaps also an increased incidence of atherosclerosis [34, 35]. Aortic distensibility correlated with serum IGF-I concentrations in 38 healthy adults [36] and a decreased aortic distensibility was demonstrated in GH-deficient patients by a Doppler ultrasound technique [37] compared to controls. These findings are associated with an increased mortality due to cardiovascular disease in GH-deficient patients [38].

Bone

A reduced bone mineral content was demonstrated in untreated adults with GH deficiency of childhood or pubertal onset [39–42] as well as in patients with GH deficiency acquired later in life [43] explaining, at least partly, the increased prevalence of osteoporosis in GH deficiency [35].

Table 1. Data on reported pregnancy outcome in GH-deficient patients (all children were delivered by caesarian section)

Reference	Diagnosis	No.	Delivery week	Birth weight, g	Placental weight, g	Comments
Tyson et al. [45]	isolated GHD	1	35.6	2,390	580	
Cassar and Verzo [47]	isolated GHD	2	37	2,180/2,340	–	delivered twins
Smallridge et al. [48]	isolated GHD	3	38	3,090	–	one previous normal pregnancy
Merimee et al., 1982	isolated GHD	4	37	'normal'	–	
Hall et al. [46]	multiple pituitary deficiency	5	36	3,560	580	
		6	33	2,570	450	
		7	37	4,230	625	

Reproduction

Patients with GH deficiency have insufficient development of external genitalia and breasts in many cases [44].

Females. Little is known about female fertility in GH-deficient patients. However, a small number of case reports have described a relatively high frequency of abortions and miscarriages, but normal pregnancy outcome when pregnancy was successfully terminated [45–48] in GH-deficient patients (table 1).

Male. Patients with isolated GH deficiency have moderately reduced testicular volume and a reduced semen volume [49] (fig. 2).

Skin

Patients with GH deficiency often have a thin, dry skin and recently a reduced sweating, determined by the pilocarpine iontophoresis method was demonstrated in GH-deficient adults [50] as well as in patients with GH insensitivity [51]. Conversely, hyperhidrosis is a classic symptom of acromegaly. Sweating is important for the maintenance of body temperature during exercise or heat exposure. A report on unexplained hyperthermia in 3 untreated adults with GH deficiency [52] indicates that GH may indeed play a role in sweating and thermoregulation. A recent study revealed that during moderate heat exposure, 6 young adults with GH deficiency had a reduced evaporative heat loss and thereby a greater body heat storage compared to 6 matched healthy subjects [53].

Psychosocial Aspects

Follow-up studies of adults treated with GH in childhood have shown a high risk of unemployment and a low marital rate [54, 55]. Assessment of psychological well-being in GH-deficient adults revealed that they perceive themselves as having physical and psychological problems [56–58]. Patients with multiple pituitary insufficiency have impaired cardiovascular responses to stress (public speaking), probably related to insufficiencies in catecholamines or cortisol secretion rather than GH [59].

Effects of GH Treatment

Body Composition

In all studies, GH treatment unanimously has increased lean body mass and improved the muscle/fat ratio in GH-deficient adults [21, 22, 26, 27, 60–64].

Muscular Performance

GH treatment improves exercise capacity [21, 25, 26] as well as muscle strength [21, 27], but no significant changes in muscle fibre diameters or proportions of type 1 or 2 muscle fibers occurred [28, 65] (fig. 4).

Cardiovascular System

GH treatment increased cardiac output [21, 66, 67] and has been reported to decrease cholesterol and low-density lipoprotein concentrations and increased high-density lipoprotein concentrations [33, 68]. Case reports exist on GH-induced improvement in cardiac status in GH-deficient patients with cardiac cachexia [69]. GH treatment did not change blood coagubility in GH-deficient adults [70]. Thus, GH seems to affect cardiovascular function and to regulate lipoprotein metabolism in adults with GH deficiency.

Kidney Function

Glomerular filtration rate and renal blood flow was significantly increased (normalized) after GH treatment [21]. Furthermore, a significant activation of the renin-

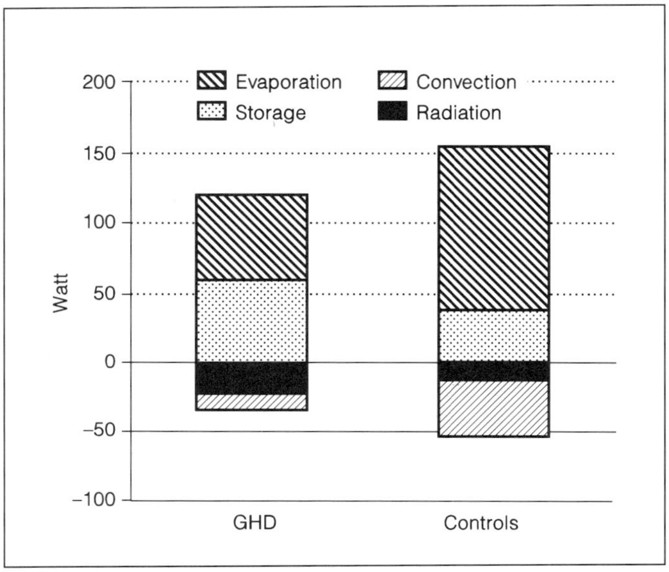

3

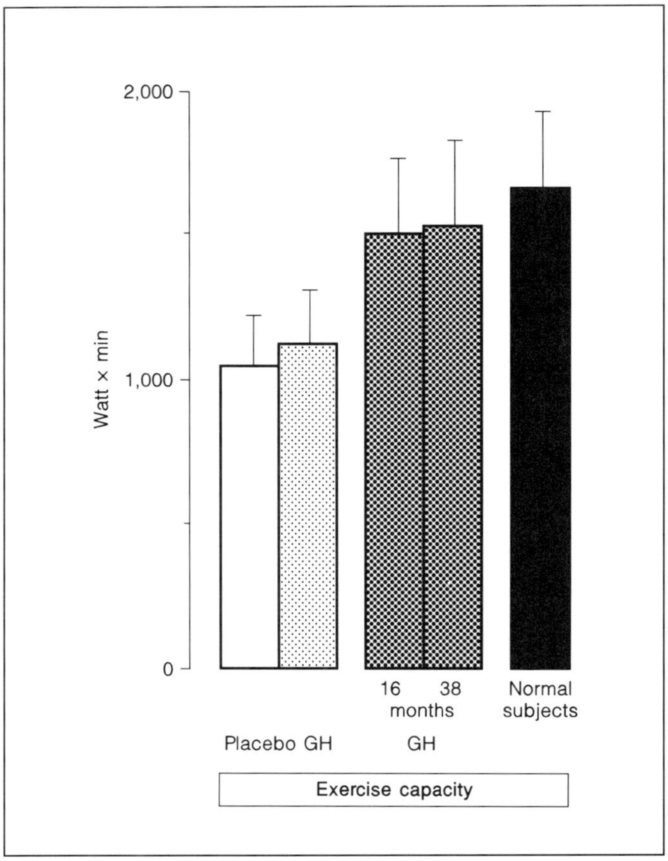

4

Fig. 3. Body heat storage, evaporation, convection and radiation during a 90-min heat exposure period (40°C) in 6 GH-deficient patients compared to 6 matched controls. Data are derived from Juul et al. [53].

Fig. 4. Exercise performance in GH-deficient patients during GH/placebo and after 38 months of GH replacement in an open trial compared to age-matched controls. The values are expressed per kg body weight. Data are derived from Jørgensen et al. [89].

angiotensin-aldosterone system has been demonstrated in GH-deficient adults [66]. This is in line with the reduction in atrial natriuretic peptide after GH treatment of normal man [71].

Thyroid Function

It is now well recognized that GH exerts important actions on peripheral iodothyronine metabolism by a stimulation of the peripheral conversion of thyroxine (T4) to triiodothyronine (T3). More interestingly, it has been demonstrated that adult GH-deficient patients exhibit subnormal T3 levels (in spite of normal T4 levels) which are normalized with GH substitution [72]. We believe that this action of GH contributes to its overall anabolic effects, as a linear correlation between T3 levels and linear growth has been demonstrated in GH-treated GH-deficient children [73].

Bone

Some controversy exists on the possible effects of GH replacement therapy on bone in the adult state. Several studies indicated increased bone turnover in GH-deficient adults during GH replacement therapy in terms of increments in serum concentrations of serum bone Gla protein, procollagens [61, 74, 75] and urinary excretion of calcium, hydroxyproline and pyrridinoline cross-links [76]. However, the effect of GH-induced increase in bone turnover on bone mineral content (BMC) is not clear. Several studies have failed to show an increase in BMC after GH treatment for up to 6 months [22, 62, 64, 75]. However, recent trials reported an increase in BMC after 12, 14 and 30 months of GH therapy, respectively [44, 63, 77, 78]. This discrepancy is probably due to the need for longer treatment periods before a beneficial effect on BMC can be seen.

Reproduction

Short-term treatment with GH did not result in any significant improvement in semen quality [49], but seemed to improve hCG-induced testosterone production [79]. The potential role of GH and IGFs on both male and female reproductive endocrinology is, however, subject to intensive research.

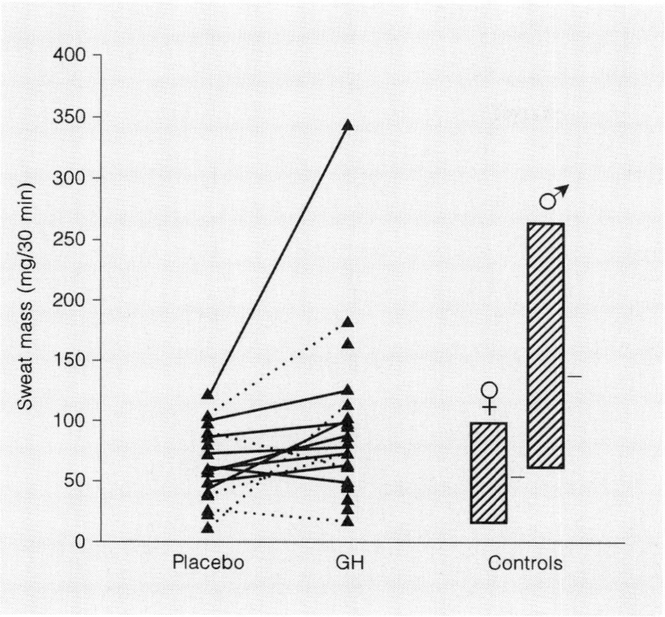

Fig. 5. Sweat secretion rate (mg/30 min) in 21 GH-deficient adults (solid lines, males; dotted lines, females) after 4 months of GH and/or placebo ($p < 0.0001$) compared to normal range for males and females [90]. Data are derived from Pedersen et al. [50].

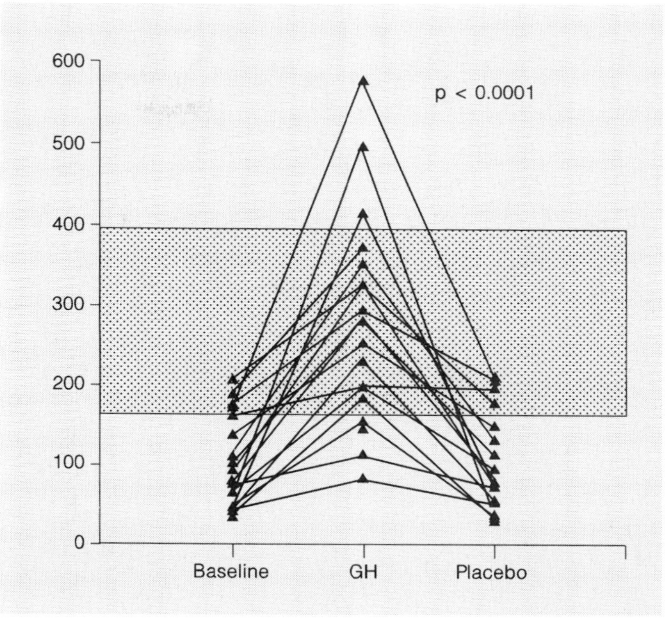

Fig. 6. Serum IGF-I levels in 21 GH-deficient young adults before and after 4 months of GH (2.0 IU/m^2) compared to placebo. The dotted area represents the 95% prediction interval for healthy adults 20–30 years of age. Data are derived from Juul et al. [44].

Skin

After 4 months of GH treatment sweating increased (normalized) significantly [50]. The possible effects of GH treatment on thermoregulation remains to be seen (fig. 5).

Psychosocial Aspects

GH treatment improves perception of energy level, mental alertness and mood [56, 80]. The mechanisms by which GH exerts its psychological effect are difficult to determine. Improved physical well-being itself might have an effect, or it could be an improved cerebral blood flow. Finally, a direct central effect of GH or IGF-I cannot be excluded. The latter hypothesis is supported by improvement of abnormal sleep pattern after GH treatment [81].

Side Effects

Clinical evidence of transient water and sodium retention after GH treatment in GH-deficient adults have been observed in most GH trials in GH-deficient adults resulting in terms of peripheral edema, body weight gain, arthralgias and symptoms of carpal tunnel syndrome. However, these side effects often disappeared spontaneously within days during continued treatment or resolved after reduction of dose. In a study of GH treatment of healthy elderly men with low IGF-I levels, a relatively high incidence of carpal tunnel symptoms and gynecomastia was demonstrated [82] relating to the posttreatment level of IGF-I. A 12-year follow-up study in a healthy population concerning the incidence of knee osteoarthritis (OA), demonstrated that the circulating IGF-I level was a strong prognostic factor for the development of knee OA. High serum concentrations of IGF-I (though within the normal range) predicted increased osteophyte formation in this otherwise healthy population [83]. In addition, a recent study demonstrated a low prevalence of OA in elderly patients with GH deficiency [84]. Therefore, one could speculate that long-term GH therapy in GH-deficient adults might cause an increased incidence of carpal tunnel syndrome and/or osteoarthritis, if IGF-I levels are increased to supraphysiological levels.

The direct metabolic actions of GH involve stimulation of lipolysis and suppression of peripheral glucose uptake. Since the prevalence of diabetes mellitus (NIDDM) is slightly elevated in acromegaly, the possible consequences of long-term GH therapy on glucose homeostasis merits attention. GH treatment was shown to reduce insulin sensitivity during a euglycemic clamp, but this effect was almost reversed following 6 months of continued treatment [85] and probably related to the reduced body fat during treatment. We have observed a completely normal response to a standard oral glucose load (75 g) following more than 3 years' GH therapy in GH-deficient adults [unpubl. data]. This suggests that the long-term insulin antagonistic effect of GH may be limited due to a concomitant normalization of body composition.

Optimization of Treatment Dose

Various GH dose regimens (from 2 to 2.8 IU/m^2) have been used [21, 60] in GH-deficient adults and this issue is presently discussed [86]. More side effects were seen the higher the dose and the older the patient. Patients have varying degrees of GH deficiency and several studies have shown that GH treatment causes a variable increase in serum IGF-I levels. As seen from figure 6, IGF-I levels during GH treatment (2.0 IU/m^2) range from below normal range to values clearly above the upper limit for age-matched healthy individuals. This implies that individualised replacement dosages according to IGF-I levels could potentially be useful. By the use of biochemical growth markers and metabolic indices, a recent study demonstrated that a replacement dose of 1–2 IU/m^2 is adequate for young adults [87]. This implies that a rough guideline for the GH replacement dose in a normal-sized individual is 1 IU/m^2/day = 0.025 IU/kg/day = 0.01 mg/kg/day = 0.175 IU/kg/week = 0.07 mg/kg/week [86]. If GH replacement therapy is supposed to mimic the physiological GH secretion, we believe that a gradual reduction in dose with increasing age would probably be advantageous or at least more physiological than maintaining 'youthful' IGF-I.

Concluding Remarks

GH deficiency in adults is associated with a number of abnormal features. GH treatment improves (normalizes) these symptoms and we believe that one should probably continue GH treatment of GH-deficient patients in adulthood after cessation of linear growth. Since data from placebo-controlled long-term follow-up studies are lacking, such patients should, however, be closely monitored and possible adverse events reported.

References

1 Ho KY, Evans WS, Blizzard RM, Veldhuis JD, Merriam GR, Samojlik E, Furlanetto R, Rogol AD, Kaiser DL, Thorner MO: Effects of sex and age on the 24-hour profile of growth hormone secretion in man: Importance of endogenous estradiol concentrations. J Clin Endocrinol Metab 1987;64:51–58.

2 Rudman D, Feller AG, Nagraj HS, Gergans GA, Lalitha PY, Goldberg AF, Schlenker RA, Cohn L, Rudman IW, Mattson DE: Effects of human growth hormone in men over 60 years old. N Engl J Med 1990;323:1–6.

3 Marcus R, Butterfield G, Holloway I: Effects of short term administration of recombinant human growth hormone to elderly people. J Clin Endocrinol Metab 1990;70:519–527.

4 Daltani MT, Pringle PJ, Hindmarsh PC, Brook CGD: What is a normal stimulated growth hormone concentration? J Endocrinol 1992;133: 447–450.

5 Rochiccioli P, Enjaume C, Tauber MT, Pienkowski C: Statistical study of 5473 results of nine pharmacological stimulation tests: A proposed weighting index. Acta Paediatr 1993;82: 245–248.

6 O'Keane V, Dinan TG: Sex steroid priming effects on growth hormone response to pyridostigmine throughout the menstrual cycle. J Clin Endocrinol Metab 1992;75:11–14.

7 Blum WF, Ranke MB, Kietzmann K, Gauggel E, Zeisel HJ, Bierich JR: A specific radioimmunoassay for the growth hormone (GH)-dependent somatomedin-binding protein: Its use for diagnosis of GH deficiency. J Clin Endocrinol Metab 1990;70:1292–1298.

8 Hasegawa Y, Hasegawa T, Yokoyama T, Kotoh S, Tsuchiya Y, Kurimoto F: Western ligand blot assay for human growth hormone-dependent insulin-like growth factor binding protein (IGFBP-3): The serum levels in patients with classical growth hormone deficiency. Endocrinol Jpn 1992;39:121–127.

9 Adan L, Souberbielle J-C, Brauner R: Diagnostic markers of permanent idiopathic growth hormone deficiency. J Clin Endocrinol Metab 1994;78:353–358.

10 Sklar C, Sarafoglou K, Whittam E: Efficacy of insulin-like growth factor binding protein 3 in predicting the growth hormone response to provocative testing in children treated with cranial irradiation. Acta Endocrinol (Copenh) 1993;129:511–515.

11 Blum WF, Albertsson-Wikland K, Rosberg S, Ranke MB: Serum levels of insulin-like growth factor I (IGF-I) and IGF binding protein 3 reflect spontaneous growth hormone secretion. J Clin Endocrinol Metab 1993;76:1610–1616.

12 Juul A, Main K, Blum WF, Lindholm J, Ranke MB, Skakkebæk NE: The ratio between serum insulin-like growth factor (IGF)-I and the IGF binding proteins (IGFBP-1, -2 and -3) decreases with age and is increased in acromegalic patients. Clin Endocrinol 1994, in press.

13 Counts DR, Gwirtsman H, Carlsson LMS, Lesem M, Cutler GB Jr: The effect of anorexia nervosa and refeeding on growth hormone-binding protein, the insulin-like growth factors (IGFs), and the IGF-binding proteins. J Clin Endocrinol Metab 1992;75:762–767.

14 Hernández M, Argente J, Navarro A, Caballo N, Barrios V, Hervás F, Polanco I: Growth in malnutrition related to gastrointestinal diseases: Coeliac disease. Horm Res 1992; 38(suppl 1):79–84.

15 Assaad SN, Cunninham GR, Samaan NA: Abnormal growth hormone dynamics in chronic liver disease do not depend on severe parenchymal disease. Metabolism 1990;39:349–356.

16 Sheppard MS, Minuk GY, Bhaumick B, Bala RM: Insulin-like growth factors (IGF) in liver disease: differential changes of IGF-I and IGF-II. Clin Invest Med 1987;10:49–53.
17 Liu F, Powell DR, Hintz RL: Characterization of insulin-like growth factor-binding proteins in human serum from patients with chronic renal failure. J Clin Endocrinol Metab 1990;70: 620–628.
18 Blum WF, Ranke MB, Kietzmann K, Tönshoff B, Mehls O: Growth hormone resistance and inhibition of somatomedin activity by excess of insulin-like growth factor binding protein in uraemia. Pediatr Nephrol 1991;5:539–544.
19 Batch JA, Baxter RC, Werther G: Abnormal regulation of insulin-like growth factor binding proteins in adolescents with insulin-dependent diabetes. J Clin Endocrinol Metab 1991;73: 964–968.
20 Miell JP, Taylor AM, Zini M, Maheshwari G, Ross RJM, Valcavi R: Effects of hypothyroidism and hyperthyroidism on insulin-like growth factors (IGFs) and growth hormone- and IGF-binding proteins. J Clin Endocrinol Metab 1993;76:950–955.
21 Jørgensen JOL, Pedersen SA, Thuesen L, Jørgensen J, Ingemann-Hansen T, Skakkebæk NE, Christiansen JS: Beneficial effects of growth hormone treatment in GH-deficient adults. Lancet 1989;i:1221–1225.
22 Binnerts A, Swart GR, Wilson JHP, Hoogerbrugge N, Pols HAP, Birkenhager JC, Lamberts SWJ: The effect of growth hormone administration in growth hormone deficient adults on bone, protein, carbohydrate and lipid homeostasis, as well as on body composition. Clin Endocrinol 1992;37:79–87.
23 De Boer H, Blok GJ, Voerman HJ, De Vries PMJM, Van der Veen EA: Body composition in adult growth hormone-deficient men, assessed by anthropometry and bioimpedance analysis. J Clin Endocrinol Metab 1992;75: 833–837.
24 Rosén T, Bosaeust I, Tölli J, Lindstedt G, Bengtsson BÅ: Increased body fat mass and decreased extracellular fluid volume in adults with growth hormone deficiency. Clin Endocrinol 1993;38:63–71.
25 Cuneo RC, Salomon F, Wiles CM, Hesp R, Sönksen PH: Growth hormone treatment in growth hormone-deficient adults. II. Effects on exercise performance. J Appl Physiol 1991;70: 695–700.
26 Jørgensen JOL, Pedersen SA, Thuesen L, Jørgensen J, Møller J, Müller J, Skakkebæk NE, Christiansen JS: Long-term growth hormone treatment in growth hormone deficient adults. Acta Endocrinol (Copenh) 1991;125:449–453.
27 Cuneo RC, Salomon F, Wiles MC, Hesp R, Sönksen PH: Growth hormone treatment in growth hormone-deficient adults. I. Effects on muscle mass and strength. J Appl Physiol 1991; 70:688–694.
28 Cuneo RC, Salomon F, Wiles CM, Round JM, Jones D, Hesp R, Sönksen PH: Histology of skeletal muscle in adults with GH deficiency: Comparison with normal muscle and response to GH treatment. Horm Res 1992;37:23–28.
29 Shahi M, Beshyah SA, Hackett D, Sharp PS, Johnston DG, Foale RA: Myocardial dysfunction in treated adult hypopituitarism: A possible explanation for increased cardiovascular mortality. Br Heart J 1992;67:92–96.
30 Johnston DG, Beshyah SA, Markussis V, Shahi M, Sharp PS, Foale RA, Skinner EM: Metabolic changes and vascular risk factors in hypopituitarism. Horm Res 1992;38(suppl 1):68–72.
31 Merola B, Cittadini A, Colao A, Longobardi S, Fazio S, Sabatini D, Saccá L, Lombardi G: Cardiac structural and functional abnormalities in adult patients with growth hormone deficiency. J Clin Endocrinol Metab 1993;77:1658–1661.
32 Rosén T, Edén S, Larson G, Wilhelmsen L, Bengtsson BÅ: Cardiovascular risk factors in adult patients with growth hormone deficiency. Acta Endocrinol (Copenh) 1993;129:195–200.
33 Cuneo RC, Salomon F, Watts GF, Hesp R, Sönksen PH: Growth hormone treatment improves serum lipids and lipoproteins in adults with growth hormone deficiency. Metabolism 1993;42:1519–1523.
34 Markussis V, Beshyah SA, Fisher C, Sharp P, Nicolaides AN, Johnston DG: Detection of premature atherosclerosis by high-resolution ultrasonography in symptom-free hypopituitary adults. Lancet 1992;340:1188–1192.
35 Wüster C, Slenczka E, Ziegler R: Erhöhte Prävalenz von Osteoporose und Arteriosklerose bei konventionell substituierter Hypophysenvorderlappeninsuffizienz: Bedarf einer zusätzlichen Wachstumshormonsubstitution? Klin Wochenschr 1991;69:769–773.
36 Hopkins KD, Lehmann ED, Gosling RG, Parker JR, Sönksen PH: Biochemical correlates of aortic distensibility in vivo in normal subjects. Clin Sci 1993;84:593–597.
37 Lehman ED, Hopkin KD, Weissberger AJ, Gosling RG, Sönksen PH: Aortic distensibility in growth hormone deficient adults. Lancet 1993;341:309.
38 Rosén T, Bengtsson B-Å: Premature mortality due to cardiovascular disease in hypopituitarism. Lancet 1990;336:285–288.
39 Hyer SL, Rodin DA, Tobias JH, Leiper A, Nussey SS: Growth hormone deficiency during puberty reduces adult bone mineral density. Arch Dis Child 1992;67:1472–1474.
40 Kaufman J-M, Taelman P, Vermeulen A, Vandeweghe M: Bone mineral status in growth hormone-deficient males with isolated and multiple pituitary deficiencies of childhood onset. J Clin Endocrinol Metab 1992;74:118–123.
41 Elgindy N, Grunditz R, Thorén M, Degerblad M, Sjöberg HE, Ringertz H: Long term follow-up of metacarpal cortical thickness and bone mineral density in panhypopituitarism. Radiol Diagn 1991;32:326–331.
42 Rosén T, Hansson T, Granhed H, Szucs J, Bengtsson B-Å: Reduced bone mineral content in adult patients with growth hormone deficiency. Acta Endocrinol (Copenh) 1993;129: 201–206.
43 Bing-You RG, Denis M-C, Rosen CJ: Low bone mineral density in adults with previous hypothalamic-pituitary tumors: Correlations with serum growth hormone responses to GH-releasing hormone, insulin-like growth factor I, and IGF binding protein 3. Calcif Tissue Int 1993;52:183–187.
44 Juul A, Pedersen SA, Sørensen S, Winkler K, Jørgensen JOL, Christiansen JS, Skakkebæk NE: Growth hormone (GH) treatment increases serum Insulin-like Growth Factor Binding Protein-3, bone isoenzyme alkaline phosphatase and forearm bone mineral content in adults with GH-deficiency of childhood onset. Eur J Endocrinol 1994, in press.
45 Tyson JE, Barnes AC, Merimee TJ, McKusick VA: Isolated growth hormone deficiency: Studies in pregnancy. J Clin Endocrinol Metab 1970;31:147–152.
46 Hall K, Enberg G, Hellem E, Lundin G, Ottosson-Seeberger A, Sara V, Trygstad O, Öfverholm U: Somatomedin levels in pregnancy: longitudinal study in healthy subjects and patients with growth hormone deficiency. J Clin Endocrinol Metab 1984;59:587–594.
47 Cassar J, Verco CJ: Successful pregnancy induced by human menopausal gonadotrophin in a patient with growth hormone deficiency and primary amenorrhoea: A case report. Br J Obstet Gynaecol 1980;87:337–339.
48 Smallridge RC, Corrigan DF, Thomason AM, Blue PW: Hypoglycemia in pregnancy. Occurrence due to adrenocorticotropic hormone and growth hormone deficiency. Arch Intern Med 1980;140:564–565.
49 Pedersen SA, Jørgensen JO, Christiansen JS, Jensen S, Jelnæs JE, Skakkebæk NE: Growth hormone and reproduction: reduced semen quality in men previously treated for growth hormone deficiency; in Frisch H, Thorner MO (eds): Hormonal Regulation of Growth. New York, Raven Press, 1989, pp 273–282.
50 Pedersen SA, Welling K, Michaelsen KF, Jørgensen JO, Christiansen JS, Skakkebæk NE: Reduced sweating in adults with growth hormone deficiency. Lancet 1989;ii:681–682.
51 Main KM, Price DA, Savage MO, Skakkebæk NE: Decreased sweating in seven patients with Laron syndrome. J Clin Endocrinol Metab 1993;77:821–823.
52 Juul A, Skakkebæk NE: Growth hormone deficiency and hyperthermia. Lancet 1991;338: 887.
53 Juul A, Behrenscheer A, Tims T, Nielsen B, Halkjær-Kristensen J, Skakkebæk NE: Impaired thermoregulation in adults with growth hormone deficiency during heat exposure and exercise. Clin Endocrinol 1993;38:237–244.
54 Dean HJ, McTaggert TL, Fish DG, Friesen HG: The educational, vocational, and marital status of growth hormone deficient adults treated with growth hormone during childhood. Am J Dis Child 1985;139:1105–1110.
55 Björk S, Jönsson B, Westphal O, Levin J-E: Quality of life of adults with growth hormone deficiency: a controlled study. Acta Paediatr Scand 1989;356(suppl):55–59.

56 McGauley GA: Quality of life assessment before and after growth hormone treatment in adults with growth hormone deficiency. Acta Paediatr Scand 1989;356(suppl):70–72.
57 Shunt SM, McKenna SP, Doward LC: Preliminary report on the development of a disease-specific instrument for assessing quality of life of adults with growth hormone deficiency. Acta Endocrinol (Copenh) 1993;128(suppl 2):37–40.
58 Rosén T, Wirén L, Wilhelmsen L, Wiklund I, Bengtsson B-Å: Decreased psychological well-being in adult patients with growth hormone deficiency. Clin Endocrinol 1994;40:111–116.
59 Stabler B, Turner JR, Girdler SS, Light KC, Underwood LE: Reactivity to stress and psychological adjustment in adults with pituitary insufficiency. Clin Endocrinol 1992;36:467–473.
60 Salomon F, Cuneo RC, Hesp R, Sönksen PH: The effects of treatment with recombinant human growth hormone on body composition and metabolism in adults with growth hormone deficiency. N Engl J Med 1989;321: 1797–1803.
61 Bengtsson B-Å, Edén S, Lönn L, Kvist H, Stokland A, Lindstedt G, Bosaeus I, Tölli J, Sjöström L, Isaksson OGP: Treatment of adults with growth hormone (GH) deficiency with recombinant human GH. J Clin Endocrinol Metab 1993;76:309–317.
62 Orme SM, Sebastian JP, Oldroyd B, Stewart SP, Grant PJ, Stickland MH, Smith MA, Belchetz PE: Comparison of measures of body composition in a trial of low dose growth hormone replacement therapy. Clin Endocrinol 1992;37:453–459.
63 Degerblad M, Elgindy N, Hall K, Sjöberg H-E, Tohrén M: Potent effect of recombinant growth hormone on bone mineral density and body composition in adults with panhypopituitarism. Acta Endocrinol (Copenh) 1992;126: 387–393.
64 Whitehead HM, Boreham C, McIlrath EM, Sheridan B, Kennedy L, Atkinson AB, Hadden DR: Growth hormone treatment of adults with growth hormone deficiency: Results of a 13-month placebo controlled cross-over study. Clin Endocrinol 1992;36:45–52.
65 Whitehead HM, Gilliland JS, Allen IV, Hadden DR: Growth hormone treatment in adults with growth hormone deficiency: Effect on muscle fibre size and proportions. Acta Paediatr Scand 1989;356(suppl):65–67.
66 Cuneo RC, Salomon F, Wilmhurst P, Byrne C, Wiles CM, Hesp R, Sönksen PH: Cardiovascular effects of growth treatment in growth-hormone-deficient adults: Stimulation of the renin-aldosterone system. Clin Sci 1991;81:587–592.
67 Amato G, Carella C, Fazio S, La Montagna G, Cittadini A, Sabatini D, Marciano-Mone C, Sacca L, Bellastella A: Body composition, bone metabolism, and heart structure and function in growth hormone (GH)-deficient adults before and after GH replacement therapy at low doses. J Clin Endocrinol Metab 1994;77:1671–1676.
68 Edén S, Wiklund O, Oscarsson J, Rosén T, Bengtsson B-Å: Growth hormone treatment of growth hormone-deficient adults result in a marked increase in Lp(a) and HDL cholesterol concentrations. Arterioscl Thromb 1993;13: 296–301.
69 Cuneo RC, Wilmhurst P, Lowy C, McGauley G, Sönksen PH: Cardiac failure responding to growth hormone. Lancet 1989;1:838–839.
70 Jørgensen JOL, Pedersen SA, Ingerslev J, Møller J, Skakkebæk NE, Christiansen JS: Growth hormone (GH) therapy in GH-deficient patients, the plasma factor VIII-von Willebrand factor complex, and capillary fragility. A double-blind, placebo-controlled cross-over study. Scand J Clin Lab Invest 1990;50:417–420.
71 Møller J, Jørgensen JOL, Møller N, Hansen KW, Pedersen EB, Christiansen JS: Expansion of the extracellular volume and suppression of atrial natriuretic peptide after growth hormone administration in normal man. J Clin Endocrinol Metab 1991;72:768–772.
72 Jørgensen JOL, Pedersen SA, Laurberg P, Weeke J, Skakkebæk NE, Christiansen JS: Effects of growth hormone therapy on thyroid function of growth hormone-deficient adults with and without concomitant thyroxine-substituted central hypothyroidism. J Clin Endocrinol Metab 1989;69:1127–1132.
73 Sato T, Suzuki Y, Takerani T, Ishiguro K, Masuyama T, Takata I, Sano M, Kawashima H, Koizumi S, Nakajima H: Enhanced peripheral conversion of thyroxine to triiodothyronine during hGH therapy in GH deficient children. J Clin Endocrinol Metab 1977;45:324–329.
74 Johansen JS, Giwercman A, Hartwell D, Nielsen CT, Price PA, Christiansen C, Skakkebæk NE: Serum bone Gla-protein as a marker of bone growth in children and adolescents: correlation with age, height, serum insulin-like growth factor I, and serum testosterone. J Clin Endocrinol Metab 1988;67:273–278.
75 Beshyah SA, Thomas E, Kyd P, Sharp P, Fariney A, Johnston DG: The effect of growth hormone replacement therapy in hypopituitary adults on calcium and bone metabolism. Clin Endocrinol 1994;40:383–391.
76 Schlemmer A, Johansen JS, Pedersen SA, Jørgensen JOL, Hassager C: The effects of growth hormone on urinary pyrridnoline cross-links in GH-deficient adults. Clin Endocrinol 1991;35: 471–476.
77 Vandeweghe M, Taelman P, Kaufman J-M: Short- and long-term effects of growth hormone treatment on bone turnover and bone mineral content in adult growth hormone-deficient males. Clin Endocrinol 1993;39:409–415.
78 O'Halloran DJ, Tsatsoulis A, Whitehouse RW, Holmes SJ, Adams JE, Shalet SM: Increased bone density after recombinant human growth hormone (GH) therapy in adults with isolated GH deficiency. J Clin Endocrinol Metab 1993; 76:1344–1348.
79 Balducci R, Toscano V, Mangiantini A, Bianchi P, Gugliemi R, Boscherini B: The effect of growth hormone administration on testicular response during gonadotropin therapy in subjects with combined gonadotropin and growth hormone deficiencies. Acta Endocrinol (Copenh) 1993;128:19–23.
80 Degerblad M, Almkvist O, Grunditz R, Hall K, Kaijser L, Knutsson E, Ringertz H, Thorén M: Physical and psychological capabilities during substitution therapy with recombinant growth hormone in adults with growth hormone deficiency. Acta Endocrinol (Copenh) 1990;123: 185–193.
81 Åstrand C, Pedersen SA, Lindholm J: The influence of growth hormone on sleep in adults with growth hormone deficiency. Clin Endocrinol 1990;33:495–500.
82 Cohn L, Feller AG, Draper MW, Rudman IW, Rudman D: Carpal tunnel syndrome and gynaecomastia during growth hormone treatment of elderly men with low circulating IGF-I concentrations. Clin Endocrinol 1993;39:417–425.
83 Schouten JS, Van-den-Oweland FA, Valkenburg HA, Lamberts SW: Insulin-like growth factor-1: a prognostic factor of knee osteoarthritis. Br J Rheumatol 1993;32:274–280.
84 Bagge E, Edén S, Rosén T, Bengtsson B-Å: The prevalence of radiographic osteoarthritis is low in elderly patients with growth hormone deficiency. Acta Endocrinol (Copenh) 1993;129: 296–300.
85 Fowelin J, Attvall S, Lager I, Bengtsson B-Å: Effects of treatment with recombinant human growth hormone on insulin sensitivity and glucose metabolism in adults with growth hormone deficiency. Metabolism 1993;42:1443–1447.
86 Ross RJM: Growth hormone replacement in adults: what dose? Clin Endocrinol 1993;39: 401–402.
87 Møller J, Jørgensen JOL, Lauersen T, Frystyk J, Næraa RW, Ørskov H, Christiansen JS: Growth hormone dose regimens in adult GH deficiency: Effects on biochemical growth markers and metabolic parameters. Clin Endocrinol 1993;39:403–408.
88 Nielsen CT, Skakkebæk NE, Richardson DW, Darling JAB, Hunter WM, Jørgensen M, Nielsen AA, Ingerslev O, Keiding N, Müller J: Onset of the release of spermatozoa (spermache) in boys in relation to age, testicular growth, pubic hair, and height. J Clin Endocrinol Metab 1986;62:532–535.
89 Jørgensen JOL, Thuesen L, Müller J, Ovesen P, Skakkebæk NE, Christiansen JS: Three years growth hormone treatment in growth hormone deficient adults: Normalization of body composition and physical performance. Acta Endocrinol (Copenh) 1994, in press.
90 Main K, Nilsson KO, Skakkebæk NE: Influence of sex and growth hormone deficiency on sweating. Scand J Clin Lab Invest 1991;51: 475–480.

Horm Res 1995;44(suppl 3):73–79

Amanda L. Ogilvy-Stuart

Department of Paediatrics, University of Oxford, John Radcliffe Hospital, Oxford, UK

Safety of Growth Hormone after Treatment of a Childhood Malignancy

Key Words
Growth hormone
Tumour recurrence
Leukaemia
Insulin-like growth factor-1

Abstract
The use of growth hormone therapy in children with radiation-induced growth hormone (GH) deficiency is widely accepted, but the safety of this mitogenic hormone, particularly in children previously treated for cancer, continues to cause concern. A variety of malignant tumours have been induced in animals exposed to supraphysiological doses of GH, whereas hypophysectomised animals appear protected from carcinogen-induced neoplasms. Growth hormone and insulin-like growth factor-1 have been shown to stimulate both proliferation and transformation of normal and leukaemic human lymphocytes in vitro when used in supraphysiological doses. Despite the theoretical arguments, there is no evidence of an increased risk of tumour recurrence following GH therapy in replacement dosage in children previously treated for a malignancy.

Introduction

The last 20 years have seen a dramatic improvement in survival rates of children treated for a large number of childhood malignancies, and attention is now being focused on the morbidity caused by treatment. Management of the two commonest childhood malignancies, acute lymphoblastic leukaemia (ALL) and brain tumours, has included cranial irradiation and hence irradiation to the hypothalamic-pituitary axis which may result in radiation-induced growth hormone (GH) deficiency, particularly with irradiation doses in excess of 27 Gy [1]. Treatment of radiation-induced GH deficiency with GH therapy is now commonplace, and with beneficial effects on physiological, metabolic and psychological parameters in GH-deficient adults [2–4] its use is likely to extend beyond the completion of growth. However, there continues to be concern about the oncogenic potential of GH used therapeutically, particularly after the successful treatment of malignancy. There are hypothetical grounds to support these fears, particularly the lymphoproliferative disorders [5].

In vitro and Animal Data

It was from studies on hypophysectomised and intact animals that a relationship between GH and tumorigenesis was first suspected. Supraphysiological doses of GH administered to female rats induced neoplastic changes in several organs; lymphosarcomas of the lung, adrenocortical and adrenomedullary carcinomas, ovarian tumours

Amanda L. Ogilvy-Stuart
Department of Paediatrics
University of Oxford
John Radcliffe Hospital
Oxford, OX3 9DU (UK)

0301–0163/95/0449–0073
$8.00/0

and breast tumours [16]. Hypophysectomised rats however had a lower incidence of carcinogen- and viral-induced neoplasms and leukaemia, which was reversed after treatment with a crude pituitary extract of 'purified' GH [7, 8]. 83% of non-hypophysectomised newborn rats developed virus-induced T cell leukaemia compared with none of the hypophysectomised rats [5]. GH administration to animals after hypophysectomy has prevented thymic involution and splenic follicular atrophy [9], and corrected reduced [^{3}H]-thymidine incorporation into DNA of thymocytes and splenocytes [10]. GH antiserum led to thymic atrophy and involution of lymphoid tissue in the spleen in mice.

Receptors for GH are present in cultured circulating human lymphocytes, and GH increases the transformation of normal human peripheral blood lymphocytes in cell culture and causes lymphocyte blastogenesis [11, 12]. GH and insulin-like growth factor-1 (IGF-1) have been shown to stimulate both proliferation and transformation of normal and leukaemic human lymphocytes in vitro when used in supraphysiological doses [5, 13, 14]. Supraphysiological doses of GH can lead to a twofold increase in colony numbers [13]. Using fresh leukaemic cells from children with ALL or acute-non-lymphoblastic leukaemia (ANLL) and recombinant GH, Estrov et al. [14] concluded that GH is mitogenic at concentrations of 150–250 ng/ml when used in combination with conditioning medium. However, GH in a concentration of less than 50 ng/ml (which is more representative of physiological concentrations) did not induce an increase in either colony numbers of human ALL cells or increase in DNA synthesis (examined by measuring [^{3}H]-thymidine incorporation) [13]. Nanogramme concentrations of human GH can potentiate T cell colony growth from normal cultures and also in a patient with a T cell variant of hairy cell leukaemia, indicating a direct rather than somatomedin-mediated role for GH in human lymphoid cells [12]. However, physiological concentrations of human GH did not influence colony formation in MOLT-4, a cell line derived from an adolescent with ALL, nor thymidine incorporation into lymphoblasts from children with ALL.

While these studies point to a role for the pituitary in exerting some effects on normally and abnormally growing cells in vitro and modulating the occurrence and growth of tumours in experimental animals, they fail to define the role of GH in this process. Hypophysectomy ablates the source of many hormones, of which GH is only one, thus suppression or regression of tumour growth in hypophysectomised animals can only suggest that GH is involved. However, it was such information that resulted in management of metastatic carcinoma by hypophysectomy. It was proposed at the time that the effects of GH were mediated through the induction of tumorigenic factors that could accelerate the growth of spontaneous tumours.

To determine whether GH has any intrinsic mitogenic action in vivo, Murphy et al. [15] examined the expression of the c-myc proto-oncogene and IGF-1 in the liver and kidney after GH administration in hypophysectomised rats. Maximal expression of c-myc mRNA levels occurred in both tissues by 1 h after a single injection of GH (100 µg/100 g body weight). Peak expression of IGF-1 mRNA transcripts by comparison occurred 6–12 h after GH administration. Furthermore, administration of recombinant IGF-1 to hypophysectomised rats (50 µg/100 g body weight) failed to increase c-myc expression in either the liver or kidney suggesting that GH itself, rather than IGF-1 initiates the mitogenic response in the liver and kidney following GH administration in hypophysectomised rats [15].

IGF-1 is thought to participate in various autocrine, paracrine and endocrine pathways to regulate both hormone function and cell proliferation. As a growth factor it probably functions as a late G1 growth signal for mesenchymal-derived cells as well as an inducer of cell proliferation and differentiation. A potent mitogenic effect of IGF-1 on a number of breast cancer cell lines has been documented [16]. Experiments with oestrogen-dependent MCF-7 cells suggest the mitogenic effects of both IGF-1 and IGF-2 are mediated via the IGF-1 receptor.

Immunoreactive IGF-1 has also been shown to be increased in human lung and colon carcinomas compared with adjacent normal tissue. Specific IGF-1 receptors have also been characterised on human T-lymphoblasts, neurogliomas and colon carcinomas.

Acromegaly

In acromegaly, in which there is a pathologically sustained high level of GH, there is an increased incidence of colonic polyps and adenocarcinomas [17–21]. There are also anecdotal reports of an association between acromegaly and other tumours: adenocarcinoma of the breast [21], adenomas and papillary carcinomas of the thyroid gland [22], parathyroid adenomas, thymomas, carcinomas of the cervix and ovary, adenocarcinomas of the uterus, renal cell carcinomas, meningiomas, neuromas, bone cancer [23] and multiple myeloma [24]. However, there is no convincing evidence of an association of specific carcinomas and acromegaly apart from colonic tumours.

Childhood Leukaemia

The scenario which has generated most concern involves the development of leukaemia de novo in children receiving GH therapy either for idiopathic GH deficiency or after treatment for a brain tumour [25–29]. A cluster was first reported from Japan in 1987 and a world-wide search for other such cases has revealed 15, 14 of whom started with GH after 1973, where leukaemia was induced 0.2–11 years after the start of treatment [30]. The doses of GH ranged from 4.5 to 18 IU/m^2/week. Seven of 14 had an additional leukaemia risk [Fanconi anaemia (n = 1), previous cyclophosphamide and azothioprine (n = 1), radiation (n = 4) and myelodysplasia (n = 1)], and a further 2 had received GH for only 2 and 3.5 months, respectively, making an aetiological link unlikely. When a link between GH treatment and leukaemia is considered on a world-wide basis, there is no striking relationship. The incidence of leukaemia in GH-treated patients is as expected in an aged-matched population in the USA, Australia, Canada and Europe (the latter only if those with additional risk factors and brief treatment periods are excluded). The Lawson Wilkins Pediatric Endocrine Society and the Human Growth Foundation of the United States convened an international workshop in 1988 [31] to review all cases of leukaemia after GH therapy, and the estimated risk of leukaemia was calculated to be 1/21,000 patient years, representing a twofold increase over the expected rate. The incidence of leukaemia in Japan after treatment with GH was calculated as 1/6,000 whilst in other countries the estimates varied between 1/12,000 and 1/35,000. An explanation for the cluster in Japan remains unexplained and may be related to factors other than GH itself [31].

More recently, leukaemia, preleukaemia or myelodysplastic syndrome has been reported in 38 patients with GH deficiency [32]. Thirty-one patients had received treatment with GH, the other 7 had not received exogenous GH treatment. Twenty-one of the 31 patients were treated for idiopathic GH deficiency. Of these, 1 had Kearns-Sayre syndrome, 1 Turner's syndrome, 2 Fanconi's anaemia and minor or multiple malformations were present in 2 patients. Of the remaining 10 patients, craniopharyngioma or radiation-induced GH deficiency was the underlying cause of the GH deficiency.

The ratios of the different leukaemia types: ALL (n = 17), ANLL (n = 9), chronic myelogenous leukaemia (n = 2), preleukaemia or myelodysplastic syndrome (n = 3) were similar to those expected in the childhood population. Eleven of these children had risk factors for the development of leukaemia.

By May 1990, the incidence of leukaemia in patients treated with GH but without additional risk factors was estimated to be 1/19,000 patient years (Lawson Wilkins Paediatric Endocrine Society Meeting, May 1990), i.e. a slightly increased risk compared with the general paediatric population. Analysing cases of ALL, ANLL, and acute and chronic myelogenous leukaemia in children treated with GH between 1988 and 1991, when ascertainment has been likely to be 100%, the incidence, 1/30,000, is comparable with the 0–15 age group.

Tumour Recurrence in Children

Few clinical data are available to assess the influence of exogenous GH therapy on tumour recurrence. Arslanian et al. [33] reported on 34 children with brain tumours sited in the pituitary region or distant from it. The predominant tumour type was craniopharyngioma (n = 18). The most common childhood brain tumour, the medulloblastoma, accounted for only 2 patients. Twenty-five children were GH deficient before treatment with surgery with or without radiotherapy, and 32 were GH deficient after treatment. Twenty-four patients received GH 0.1 IU/kg three times a week, of whom 8 had a tumour recurrence, 5 within the first 7 months of treatment. Three of 11 children who did not receive GH therapy had recurrences, including 1 who was receiving testosterone. Those with recurrences grew significantly slower than those without recurrence during the 1st year of GH therapy. Though these results suggested that GH was not influencing tumour recurrence, concern was raised by the close temporal relationship between recurrence and initiation of GH in 5 cases.

In a similar study, Rodens et al. [34] reviewed tumour recurrence in 80 brain tumour patients. Again 50% had a craniopharyngioma. Five of 31 patient who received GH therapy had tumour recurrences compared with 18 recurrences in 31 GH-deficient patients who did not receive GH therapy. A third of 18 patients who were not GH deficient developed a tumour recurrence. In these two studies it is difficult to extrapolate from treatment and control groups comprised of heterogeneous tumour types with different growth and recurrence rates. This was addressed by Clayton et al. [35] who studied tumour recurrence rates in 24 children with brain tumours distant from the hypothalamic-pituitary axis and 6 with ALL who had received GH therapy. The recurrence rates in those with medulloblastoma treated with radiotherapy in the North West region of England between 1972 and 1982 who had survived more than 2.5 years in those treated with GH and those who did not receive GH were compared. In those with a diagnosis of medulloblastoma, 2 of 11 (18%) treated with GH relapsed, compared with 3 of 17 (18%) not treated with GH. Only 8 patients with an ependymoma were irradiated between 1972 and 1982 and 2 were treated with GH. Four relapsed within 1.5 years. The 2 treated with GH had late recurrences 3.5 and 4.1 years after diagnosis. The only other child with a brain tumour to relapse had a low-grade glioma and relapsed more than 12 years after diagnosis and after more than 3 years of GH. Late recurrences are known to occur even in low-grade gliomas. None of the children with ALL treated with GH relapsed. The findings suggested GH therapy does not increase tumour recurrence rates in medulloblastoma, glioma and ALL. However, the numbers were small and no formal statistical analysis was applied.

An analysis of deaths in recipients of pituitary GH in the UK cited brain tumour recurrence as one of the most common causes, but there was no comparative group who had not received GH to complement the study population. Therefore it could not be established whether the use of GH was contributory to tumour recurrence [36].

A study of 6,284 recipients of pituitary GH in the United States between 1963 and 1985 sought to determine the risk of developing leukaemia or lymphoma [37]. Eighty-four percent of the cohort were contacted, and a positive history of malignancy confirmed by review of clinical records. Deaths in the cohort were monitored through the National Death Index, and medical records and/or death certificates were reviewed in 94% of those who had died. The denominator for expected number of cases of leukaemia or lymphoma for an age-, sex- and race-matched population was taken from the Surveillance, Epidemiology and End Results programme between 1982 and 1986 (a period when 44% of the defined person-years of risk occurred).

There were 6 cases of leukaemia. Four patients had been treated for craniopharyngioma and 3 of these had received cranial irradiation. One patient had received previous cranial irradiation for an astrocytoma and 1 patient with fetal alcohol syndrome was treated with GH as part of the therapy of panhypopituitarism. Three of these cases had been ascertained from the original analysis and 3 collected from passive reporting of occurrence of leukaemia or detection through the National Death Index in the 2.5 years after the original contact and interviewing period and hence may represent an underrepresentation in cases of leukaemia, particularly if non-fatal. There were an estimated 83,917 person-years of risk and an expected 2.26 cases of leukaemia which gave an observed/expected ratio of 2.6:1 (95% confidence interval 1.2–5.2). The patients who developed leukaemia had received a mean of 3.5 years of GH whereas the mean duration in the entire cohort was 2.9 years.

Five patients developed lymphoma, 1 had been treated for a craniopharyngioma without radiotherapy, 3 had idiopathic GH deficiency and 1 Turner's syndrome. The 5 cases did not represent a significant excess in the expected (4.5) number of cases.

The association of leukaemia following treatment of craniopharyngioma has not been reported elsewhere. Clayton et al. [38] found no association between the use of GH after treatment of craniopharyngioma and tumour recurrence and did not report any cases of leukaemia in 23 children who had received GH.

To determine whether the use of GH in the treatment of radiation-induced GH causes tumour recurrence, we have analysed our tumour recurrence rates (i.e. recurrence of the original tumour either at the primary site or elsewhere) between children treated with GH for radiation-induced GH deficiency and an untreated population [39].

Children treated with GH following treatment for a brain tumour often have abnormal central nervous system radiology at the outset, so to determine if such children are at special risk for a clinically apparent relapse following GH administration, the CT scans in the children with brain tumours treated with GH were reviewed at the time of GH commencement and subsequently.

The population studied was all children who were clinically relapse free at least 2 years after diagnosis of a brain tumour (n = 207) between 1965 and 1989, or ALL (n = 161) between 1970 and 1989 in the North-West region of England. The records of all children registered with the North-West Children's Cancer Registry were reviewed and diagnosis, treatment details (use of radiotherapy and chemotherapy), relapse-free survival, age at diagnosis and sex were noted. Sixty-eight children, 53 with a brain tumour distant from the hypothalamic-pituitary axis (36 boys) and 15 with ALL (9 boys) were treated with GH for radiation-induced GH deficiency. GH therapy was commenced at least 2 years after completion of radiotherapy (i.e. after the time when tumour recurrence is most likely

Table 1. Clinical details of children treated with GH (individual data on children who relapsed and mean data on those who did not)

Patient	Tumour	Age at radio-therapy	Time from from radio-therapy to GH, years	GH treatment years	Time to relapse from GH commence-ment, years	Time since GH stopped years
1	M	13.1	2.9	3.0	3.4	1.8
2	M	8.3	3.4	0.7	0.7	–[a]
3	E	10.9	2.4	0.4	0.4	–[a]
4	E	11.3	2.5	1.0	5.3	4.4
5	A	2.1	9.1	3.3	3.3	12.4[a]
6	ALL	3.0	8.4	1.7	1.7	2.4[a,b]
Others						
GH completed		5.4	5.9	4.1	–	5.1 (n = 27)
GH ongoing		4.0	5.0	2.8	–	–

M = Medulloblastoma; E = ependymoma; A = astrocytoma.
[a] Relapse on GH.
[b] Previous relapse before GH treatment.

to occur). The dose of GH was 12 IU/week before 1989 and 0.5 IU/kg/week after 1989.

For those children with a diagnosis of a brain tumour, Cox's regression model was employed to compare children who did or did not receive GH at the time of each recurrence allowing for other covariates thought to influence relapse-free survival (diagnosis, sex, age at diagnosis and whether or not chemotherapy was included in the primary treatment).

Excluded from the analysis were children who had relapsed within 2 years from diagnosis, those that had not received radiotherapy during their primary treatment (and who would therefore not be at risk of radiation-induced GH deficiency), 6 children who received GH therapy within 2 years from diagnosis, and 2 children in whom treatment details were incomplete.

Five children treated with GH have had a clinical relapse associated with recurrence of their brain tumour. In 2 this occurred after completion of GH therapy. The patient with an astrocytoma has survived the recurrence. Forty-two children who did not receive GH have relapsed. None of the children who received GH therapy but were subsequently removed from the analysis has relapsed. The age at irradiation, time and length of GH treatment and years of survival since completion of GH therapy in all GH-treated children with and without a tumour recurrence is shown in table 1.

Adjusting for variables other than GH which might affect tumour recurrence, the estimated relative risk of tumour recurrence was 0.82 (95% confidence interval: 0.28–2.37). In each tumour category, there was no association between the use of GH and subsequent tumour recurrence.

Forty-four children with brain tumours had CT scans around the time of GH commencement. In addition to postoperative and radiotherapy changes, 19 of these scans were abnormal. Ten were described as showing residual tumour, and the remainder had non-enhancing low attenuation or cystic lesions in the area of the original tumour. The 5 children who had relapsed had CT scans at the time of GH commencement. One had evidence of residual tumour, and 1 had a low density non-enhancing lesion. In all 5 children, subsequent relapse was confirmed by an enhancing tumour on CT scan.

The scan appearance at follow-up of 14 of the 39 children who clinically remain relapse free is described in figure 1. In none has there been a deterioration in appearance of the residual lesion.

For the children with a primary diagnosis of ALL we have documented the relapse numbers of those who were not treated with GH after 5 years of relapse-free survival.

484 cases of ALL were diagnosed between 1970 and 1989 in the North-West region. 161 children were relapse free at 5 years, and 14 received GH. One additional child who received GH has relapsed. He had had two previous relapses but was thought to have a reasonable prognosis at the time of GH commencement. No child who had been relapse free at the time of GH commencement has subsequently relapsed. Eleven of 147 (7%) children not receiving GH have relapsed and a further 3 died from other related

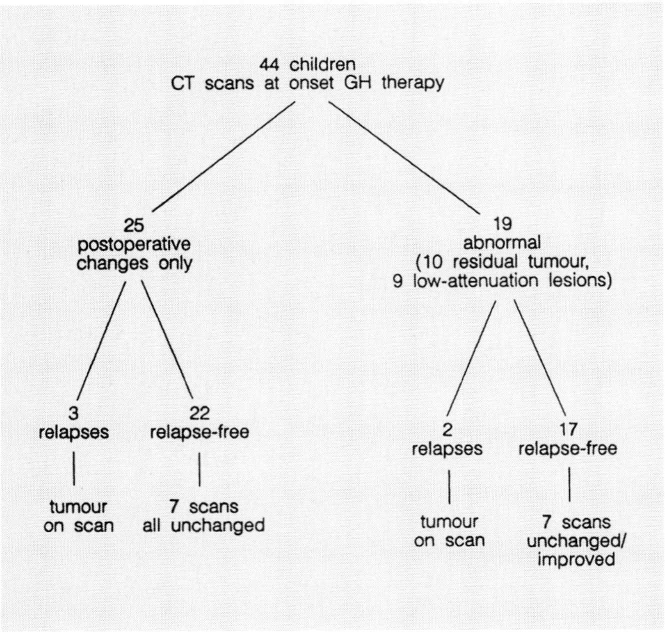

Fig. 1. Computed tomograph appearances.

causes. No child receiving GH has developed a second primary tumour and no child treated for a brain tumour and subsequently with GH has developed leukaemia.

In conclusion, in our population there is no evidence of an increased risk of tumour recurrence following GH therapy, however because of the confidence interval range continued surveillance is essential. An abnormal CT scan is not a contraindication to GH therapy. Studies worldwide do not demonstrate an increased risk of development of leukaemia in those without additional risk factors. In patients with an underlying predisposition (such as chromosome fragility or previous exposure to cytotoxic chemotherapy or irradiation) the risk of leukaemia may be increased up to 2.5-fold compared with the general population. The association of leukaemia with craniopharyngioma requires further evaluation.

Acknowledgements

Thanks to Prof. Stephen Shalet for his helpful comments.

References

1 Clayton PE, Shalet SM: Dose dependency of onset of radiation-induced growth hormone deficiency. J Pediatr 1991;118:226–228.
2 Salomon F, Cuneo RC, Hesp R, Sonkson PH: The effects of treatment with recombinant human growth hormone on body composition and metabolism in adults with growth hormone deficiency. N Engl J Med 1989;321:1797–1803.
3 Jorgensen JOL, Pedersen SA, Thuesen L, Jorgensen J, Ingemann-Hansen T, Skakkebaek NE, Christiansen JS: Beneficial effects of growth hormone treatment in GH-deficient adults. Lancet 1989;ii:1221–1225.
4 Holmes SJ, Whitehouse RW, Swindell R, Economou G, Adams JE, Shalet SM: Effect of growth hormone replacement on bone mass in adults with adult onset growth hormone deficiency. Clin Endocrinol (Oxf) 1995;42:627–633.
5 Rogers PC, Komp D, Rogol A, Sabio H: Possible effect of growth hormone on development of acute lymphoblastic leukaemia. Lancet 1977;ii:434–435.
6 Moon HD, Simpson ME, Li CH, Evans HM: Neoplasms in rats treated with pituitary growth hormone. I. Pulmonary and lymphatic tissues. Cancer Res 1950;10:297–308.
7 Moon HD, Li CH, Simpson ME: Effect of pituitary hormones on carcinogenesis with 9,10-dimethyl-1,2-dibenzanthracene in hypophysectomized rats. Cancer Res 1956;16:111–116.
8 Bentley HB, Hughes ER, Peterson RDA: Effect of hypophysectomy on a virus-induced T-cell leukaemia. Nature 1974;252:747–748.
9 Talwar GP, Pandian MR, Kumar N, Hanjan SNS, Saxena RK, Krishnaraj R, Gupta SL: Mechanism of action of pituitary growth hormone. Recent Prog Horm Res 1975;31:141–174.
10 Talwar GP, Hanjan SNS, Kidwai S, Gupta PD, Mehrota NN, Saxena R, Bhattarai Q: Growth hormone action on thymus and lymphoid cells; in Pecile A, Muller EE: Growth Hormone and Related Peptides. Proc 3rd Int Symp, Milan, 1975, pp 104–115.
11 Astaldi A, Yalcin B, Meardi G, Burgio GR, Merolla R, Astaldi G: Effect of growth hormone on lymphocyte transformation in cell culture. Blut 1973;26:74–81.
12 Mercola KE, Cline MJ, Golde DW: Growth hormone stimulation of normal and leukemic human T-lymphocyte proliferation in vitro. Blood 1981;58:337–340.
13 Blatt J, Wenger S, Stitely S, Lee PA: Lack of mitogenic effects of growth hormone on human leukemic lymphoblasts. Eur J Pediatr 1987;146:257–260.
14 Estrov Z, Meir R, Barak Y, Zaizov R, Zadik Z: Human growth hormone and insulin-like growth factor-1 enhance the proliferation of human leukaemic cells. J Clin Oncol 1991;9:394–399.
15 Murphy LJ, Bell GI, Friesen HG: Growth hormone stimulates sequential induction of c-myc and insulin-like growth factor 1 expression in vivo. Endocrinology 1987;120:1806–1812.
16 Cullen KJ, Yee D, Sly WS, Perdue J, Hampton B, Lippman ME, Rosen N: Insulin-like growth factor expression and function in human breast cancer. Cancer Res 1990;50:48–53.
17 Ezzat S, Strom C, Melmed S: Colon polyps in acromegaly. Ann Intern Med 1991;114:754–755.
18 Wright AD, Hill DM, Lowy C, Fraser TR: Mortality in acromegaly. QJ Med 1970;39:1–16.
19 Brunner JE, Johnson CC, Zafar S, Peterson EL, Brunner JF, Mellinger RC: Colon cancer and polyps in acromegaly: Increased risk associated with family history of colon cancer. Clin Endocrinol 1990;32:65–71.
20 Ron E, Gridley G, Hrubec Z, Page W, Arora S, Fraumeni JF: Acromegaly and gastrointestinal cancer. Cancer 1991;68:1673–1677.
21 Pines A, Rosen P, Ron E, Gilat T: Gastrointestinal tumors in acromegalic patients. Am J Gastroenterol 1985;80:266–269.
22 Barzilay J, Heatley GJ, Cushing GW: Benign and malignant tumors in patients with acromegaly. Arch Intern Med 1991;151:1629–1632.
23 Ratner RE, Hare JW: Association of acromegaly and chondrosarcoma. South Med J 1983;76:1181–1182.

24 Hagg E, Asplund K, Holm J: Acromegaly and multiple myeloma (letter). Ann Intern Med 1988;109:437–438.
25 Delemarre-Van de Waal HA, Ondink JH, de Grauw TJ, de Waal FC: Leukaemia in patients treated with growth hormone. Lancet 1988;i: 1159.
26 Watanabe S, Tsunematsu Y, Fujimoto J, Komiyama A: Leukaemia in patients treated with growth hormone. Lancet 1988;i:1159.
27 Frisch H, Thun-Hohenstein L, Balzar E: Leukaemia and growth hormone (letter). Lancet 1988;i:1335.
28 Blatt J, Penchansy L, Phebus C: Leukaemia in a child with a history of medulloblastoma. Pediatr Hematol Oncol 1991;8:77–82.
29 Baker DL, Kees UR, Price PJ, Willoghby MLN: Late leukemic relapse 10 years from diagnosis in a child on recombinant human growth hormone. Pediatr Hematol Oncol 1993; 10:55–62.
30 Stahnke N, Zeisel HJ: Growth hormone therapy and leukaemia. Eur J Pediatr 1989;148: 591–596.
31 Fisher DA, Job JC, Preece MA, Underwood LE: Leukaemia in patients treated with growth hormone. Lancet 1988;i:1159–1160.
32 Shalet SM: Leukaemia in children treated with growth hormone. J Pediatr Endocrinol 1993;6: 109–111.
33 Arslanian SA, Becker DJ, Lee PA, Drash LA, Foley TP: Growth hormone and tumor recurrence. Findings in children with brain neoplasms and hypopituitarism. Am J Dis Child 1985;139:347–350.
34 Rodens KP, Kaplan SL, Grumbach MM, Teller WM: Does growth hormone therapy increase the frequency of tumor recurrence in children with brain tumors? Acta Endocrinol suppl (Copenh) 1987;283:188–189.
35 Clayton PE, Shalet SM, Gattamaneni HR, Price DA: Does growth hormone cause relapse of brain tumours? Lancet 1987;i:711–713.
36 Buchanan CR, Preece MA, Milner RDG: Mortality, neoplasia, and Creutzfeldt-Jakob disease in patients treated with human pituitary growth hormone in the United Kingdom. BMJ 1991;302:824–828.
37 Fradkin JE, Mills JL, Schonberger LB, Wysowski DK, Thomson R, Durako SJ, Robison LL: Risk of leukemia after treatment with pituitary growth hormone. JAMA 1993;270:2829–2832.
38 Clayton PE, Price DA, Shalet SM, Gattemaneni HR: Craniopharyngioma recurrence and growth hormone therapy (letter). Lancet 1988; i:642.
39 Ogilvy-Stuart AL, Ryder WDJ, Gattamaneni HR, Clayton PE, Shalet SM: Growth hormone and tumour recurrence. BMJ 1992;304:1601–1605.

Author Index

Subject Index

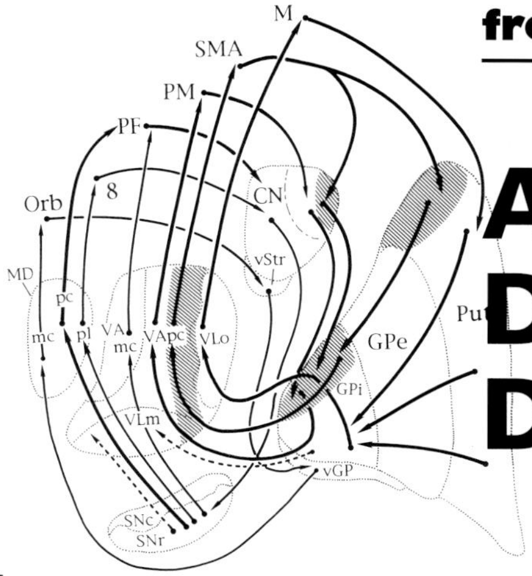

Stimulating insights from painstaking research

Age-Related Dopamine-Dependent Disorders

Editors:
Masaya Segawa
Yoshiko Nomura

The intriguing relationship between the age and sex of a patient and the nature and course of the development of a basal ganglia disorder is examined in this volume. To clarify the reasons behind this connection, long recognized but little researched, the authors have evaluated the information gleaned from the results of experimental and molecular biological studies, stereotaxic surgery and neurological research. Highly specialized articles discuss aspects of three movement disorders: hereditary progressive and dopa-responsive dystonia, juvenile Parkinsonism and Tourette's syndrome. Of particular interest is the newly discovered information about the genetic mechanism underlying the age-sex link in the development of these diseases.
Neurologists, geneticists and scientists doing research in the field of movement disorders will find this book a valuable source of new information.

Contents

Pathophysiology and Molecular Biology of Dopa-Related Disorders in Childhood and Adolescence
Hereditary Progressive Dystonia with Marked Diurnal Fluctuation and Dopa-Responsive Dystonia: Pathognomonic Clinical Features • Juvenile Parkinsonism and Other Dopa-Responsive Syndromes • Phenotypic Polymorphism of Dopa-Responsive Dystonia in Russia • The Pathogenesis of Tourette's Syndrome: Role of Biogenic Amines and Sexually Dimorphic Systems Active in Early CNS Development • Pathophysiology of Muscle Tone and Involuntary Movement in Young-Onset Basal Ganglia Disorders • Voluntary Saccades in Normal and in Dopamine-Deficient Subjects • Phasic Activity during REM Sleep in Movement Disorders • PET Studies of the Dopaminergic and Opioid Function in Dopa-Responsive Dystonic Syndromes and in Tourette's Syndrome • Fluorodopa PET Scans of Juvenile Parkinsonism with Prominent Dystonia in Relation to Dopa-Responsive Dystonia • Striatal Dopamine D_2 Receptors in Dopa-Responsive Dystonia and Parkinson's Disease • Striatal Dopamine in Dopa-Responsive Dystonia: Comparison with Idiopathic Parkinson's Disease and Other Dopamine-Dependent Disorders • Dopa-Responsive Dystonia: Clinical, Pathological, and Genetic Distinction from Juvenile Parkinsonism • The Gene for Hereditary Progressive Dystonia with Marked Diurnal Fluctuation Maps to Chromosome 14q • Early-Onset, Generalized Dystonia Caused by DYT1 Gene on Chromosome 9q34

Neuronal Circuits and Compartments of the Basal Ganglia, and Their Clinical Manifestations
Some Aspects of Basal Ganglia-Thalamocortical Circuitry and Descending Outputs of the Basal Ganglia • Dopaminergic Innervation of Primate Cerebral Cortex. An Immunohistochemical Study in the Japanese Macaque • Interactions between Glutamate and Dopamine in the Ventral Striatum: Evidence for a Dual Glutamatergic Function with Respect to Motor Control • Locomotion and Posture: Modulation by Upper Neurons • Function of the Indirect Pathway in the Basal Ganglia Oculomotor System: Visuo-Oculomotor Activities of External Pallidum Neurons • Role of the Nigrostriatal Dopamine System in Corticostriatal Signal Transmission in Alert Animals • The Role of the Descending Pallido-Reticular Pathway in Movement Disorders • Experiences from Human Stereotaxic Surgery

Monoamine Neurons: Gene and Gender Differentiation
Catecholaminergic Systems and the Sexual Differentiation of the Brain. Cellular Mechanisms and Clinical Implications • Embryonic Striatal Grafting: Progress and Future Directions for Therapeutic Approaches to Neurodegenerative Diseases of the Basal Ganglia • Basic Fibroblast Growth Factor in the Substantia nigra in Parkinson's Disease and Normal Aging

Monographs in Neural Sciences, Vol. 14
Series Editor: Korczyn, A.D. (Tel Aviv)
ISSN 0300–5186

Fields of Interest: Neurology; Pediatrics; Genetics; Biochemistry, Molecular Biology, Physiology

Age-Related Dopamine-Dependent Disorders
International Symposium on Age-Related Monoamine-Dependent Disorders and Their Modulation by Gene and Gender, Tokyo, November 1993
Editors: Segawa, M.; Nomura, Y. (Tokyo)
X + 254 p., 72 fig., 19 tab., hard cover, 1995
SFr. 240.– / DM 287.– / US$ 208.75
Prices subject to change
DM price for Germany, US$ price for USA only
ISBN 3–8055–5960–7

KI 95514

KARGER

Horm Res
ISSN 0301–0163
HRMRA3
44(suppl 3) 1–80 (1995)

Hormone Research

Growth and Growth Disorders: An Update

Diagnostic and Therapeutic Aspects

Since the availability of unlimited amounts of recombinant human growth hormone (rhGH), rapid progress has been made in the field of growth delay. This development has enabled the use of rhGH not only for the treatment of GH-deficient children, but also for children of short stature not necessarily affected by GH deficiency and even for patients with other disorders who may benefit from the anabolic action of GH.

This timely publication provides an authoritative review of the latest research advances and discusses questions and problems that have arisen owing to changes in the therapeutic applications of rhGH. One such issue concerns the safety and metabolic consequences of supraphysiological doses of rhGH in non-GH-deficient children. The real benefit of this expensive therapy, which compels the child to daily injections, is also assessed. This special issue of Hormone Research is a valuable update for pediatricians and endocrinologists involved in the clinical management of children with growth disorders.

KARGER